Managing and Engineering in Complex Situations

TOPICS IN SAFETY, RISK, RELIABILITY AND QUALITY

Volume 21

Aims and Scope. Fundamental questions which are being asked these days of all products, processes and services with ever increasing frequency are:

What is the risk?
How safe is it?
How reliable is it?
How good is the quality?
How much does it cost?

This is particularly true as the government, industry, public, customers and society become increasingly informed and articulate.

In practice none of the three topics can be considered in isolation as they all interact and interrelate in very complex and subtle ways and require a range of disciplines for their description and application; they encompass the social, engineering and physical sciences and quantitative disciplines including mathematics, probability theory and statistics.

The major objective of the series is to provide series of authoritative texts suitable for academic taught courses, reference purposes, postgraduate and other research and practitioners generally working or strongly associated with areas such as:

Safety Assessment and Management
Emergency Planning
Risk Management
Reliability Analysis and Assessment
Vulnerability Assessment and Management
Quality Assurance and Management

Special emphasis is placed on texts with regard to readability, relevance, clarity, applicability, rigor and generally sound quantitative content.

For further volumes:
http://www.springer.com/series/6653

Samuel F. Kovacic • Andres Sousa-Poza
Editors

Managing and Engineering in Complex Situations

Springer

Editors
Samuel F. Kovacic
Batten College of Engineering and Technology
Department of Engineering Management and Systems Engineering
Old Dominion University
Norfolk, VA, USA

Andres Sousa-Poza
Batten College of Engineering and Technology
Department of Engineering Management and Systems Engineering
Old Dominion University
Norfolk, VA, USA

ISSN 1566-0443
ISBN 978-94-007-5514-7 ISBN 978-94-007-5515-4 (eBook)
DOI 10.1007/978-94-007-5515-4
Springer Dordrecht Heidelberg New York London

Library of Congress Control Number: 2012954053

Printed on acid-free paper

Springer is part of Springer Science+Business Media (www.springer.com)

Contents

Chapter 1
Introduction and Overview

Andres Sousa-Poza

Abstract A brief introduction of complex situations and wicked problems is presented to frame the purpose and topics of the book. Some of the more salient features of working within complex situations are highlighted. The criticality of understanding the nature of the situation and establishing approaches and methods that are consistent with those conditions is established. Finally, an overview of the different parts and chapters included in the book is provided. The book is divided into three parts: part I includes theory of complex situations and wicked problems, part II discusses features of organizing, managing and engineering in wicked problems, and finally, part III presents several cases and approaches related to wicked problems.

Keywords Complex situations • Complexity • Wicked problems • Complex systems • MECS forum

1.1 Purpose of the Book

1.1.1 Setting the Stage: Introducing Wicked Problems

This book is driven by an indelible need to gain a greater understanding of wicked problems. Investigation into so appropriately named problems is not new. The first documented use of the term dates back to 1967 by Churchman (Churchman 1967)

A. Sousa-Poza, Ph.D. (✉)
Batten College of Engineering and Technology, Department of Engineering Management and Systems Engineering, Old Dominion University,
Kaufman Hall 242, 23529, Norfolk, VA, USA
e-mail: asousapo@odu.edu

S.F. Kovacic and A. Sousa-Poza (eds.), *Managing and Engineering in Complex Situations*, Topics in Safety, Risk, Reliability and Quality 21, DOI 10.1007/978-94-007-5515-4_1,

who attributes the term to Horst Rittel. Rittel and Weber followed this up in 1973 with a treatise on the "Dilemmas in a General Theory of Planning" in which they discussed the impact of Wicked Problems. They highlighted ten "rules" that describe the manner in which "wickedness" impacts planning. These are:

1. There is no definitive formulation of a wicked problem.
2. Wicked problems have no stopping rule.
3. Solutions to wicked problems are not true-or-false, but good-or-bad.
4. There is no immediate and no ultimate test of a solution to a wicked problem.
5. Every solution to a wicked problem is a "one-shot operation"; because there is no opportunity to learn by trial-and-error, every attempt counts significantly.
6. Wicked problems do not have an enumerable (or an exhaustively describable) set of potential solutions, nor is there a well-described set of permissible operations that may be incorporated into the plan.
7. Every wicked problem is essentially unique.
8. Every wicked problem can be considered to be a symptom of another problem.
9. The existence of a discrepancy in representing a wicked problem can be explained in numerous ways. The choice of explanation determines the nature of the problem's resolution.
10. The planner (designer) has no right to be wrong.

These rules still provide a great insight into some of the more pressing difficulties of dealing with wicked problems. More importantly was the absoluteness with which the rules are stated. For example, "Wicked Problems have no stopping rule", does not imply that there may be difficulties in determining a sufficient and necessary outcome. Rather it stipulates that wicked problems are persistent in nature; that they do not have a clear end.

Interestingly, it leaves a choice to be made if we wish to be consistent in our thinking.

(a) If we buy into this attribute of a wicked problem, then any method or approach that we would like to use must allow for this attribute. Methods cannot be used that require the establishment of clearly articulated end points whether they are predefined outcomes or deadlines.
(b) Alternatively, if we do not buy into the attribute of a wicked problem, we are left with the option of using methods that would not be conditionally constrained, but would have to deal with the consequence of having disregarded the attribute should it come to pass that Rittel and Weber (1973) framed these sensibly.

This leads to an important proposition in dealing with any problem. Behaving in a manner that is congruent with a situation is something that we accept as desirable. Not that the choice of the best method or approach will guarantee success (since there are several other rules that make this implausible); but rather, if we use a method that is not appropriate for specific set of conditions, we might well presume a negative outcome or failure. Consequently, we also accept that it is inconsistent to act in a manner that is incongruent with the conditions surrounding a problem.

This basic understanding is not necessarily novel. Wicked problems, even though they were not referred to as such, have a rich history in literature. Problems of an ethical nature have been at the core of written work for many millennia, as have works related to leadership, government and engineering. For example:

- Sun Tzu (2005) in the Art of War recounts ways, heuristics, to help deal with the uncertainties related to warfare.
- Machiavelli (1961) in the Prince focused on governance in medieval Europe.
- Tolstoy (1998) in War and Peace expounds on the transient nature of warfare as well as the importance of having a contextually rich understanding of a problem if one needs to make decisions. Tolstoy goes on to exemplify the difficulties that arise if these "rules" are not taken into consideration, questioning the wisdom of having the Tsar making military decisions

Some of the more pervasive discussions on the attributes of complex situations date back to the earliest writing. Ancient Greek literature is beset by debates on the intricacies of ethics, morality of decisions, perception, and human motivation. They also extensively debated attributes that form the cornerstone of wicked problems to this day.

- Zeno (In Plato 370 BC), for example, extensively debated the distinctions between space and time highlighting that once we focus on one we loose the ability to see the other. This is analogous to the difficulties that would be encountered in planning. Even though we know what it is that we intend to achieve, the transition between states is not always apparent.
- Documented debates on dualism started with philosophers such as Plato (Hart 1996). Many of these led to the mind-body duality that was formalized by Descartes (Robinson 2003). The basic separation between particulate matter and a superseding (non-physical) construct is however the basis of many discussions on organizational theory differentiating for example rational and natural definitions of an organization (Scott 2002).
- Teleological dilemmas and distinctions still underscore concepts such as mission and vision in organizational studies and decision making in management sciences. Democritus argued that the reason for an entity's state is the make-up and configuration of its elements. This is very consistent with modern systems thinking where the parts define the whole and the behavior and purpose is emergent from the behavior and interaction of those parts. An acronym POSIWID (Beer 2002); the purpose of a system is what it does, underscores this sentiment. From an Aristotelian point of view, everything, from engine to organization, is defined by its purpose or final cause. For sufficiency to exist, there must be a *reason* for why something is as it is. The make-up and configuration of its parts are driven by this reason. Discussions on teleology persist in management literature, for example in Stacey's (Stacey et al. 2000) work on Management and Complexity.

1.1.2 *Setting the Stage: Importance of Understanding Wicked Problems*

There cannot be any doubt that the topic is persistent. It is however also pervasive.

Wicked problems in some manner or other have always been present. To say that things have now only become complex would be incorrect in a general sense, even though it is true for some specific cases. For example, the problems that have arisen due to the interconnectedness provided by modern information and communications technologies were obviously not something that had to be considered in the past (Zhang 2002; Krell 2000). It is not even necessarily true that they have become more complex, after all, an inability or difficulty to communicate presents its own challenges. Rather the source of the wickedness may have changed, changing the nature of the problems being faced. At the same time however, the problem being discussed seem to be very similar to problems of the past. There is intransience in the very attributes of complex situations or wicked problems even thought the problems have changed.

The increasing number of groups and studies that are focusing on complexity in some form is another indicator of the pervasiveness of wicked problems, and at least an increasing level of recognition that there is a set of problems that are not well dealt with using the means that we presently have at our disposal.

For many, the pervasiveness of these problems establishes their importance. There is however a further level of criticality that must be explored; namely, the consequence of not gaining a better understanding of wicked problems, be it as complex situations or through other means.

It is more difficult to establish the consequences of behaviors that are incongruent with a problem, since this misalignment is seldom considered and therefore not identified and reported. Mostly failure studies do not reach back and question the formulation of the problem and its consequence, but rather evaluate within the construct of the formulation that was applied. Even so, when one compiles some of the data available, it becomes clear that the consequence of failure is extremely high. This is sufficiently so to warrant understanding the problem better even if no direct link can be made between the failures and the "wickedness" of the problem. It is the degree of the findings of the failure studies that puts into question the possibility that failure was merely due to first order (ineffective implementation of methods) or second order (lack of control) effect. When typical performance measures are orders of magnitude off target, it should become necessary to question the thought processes that generated the measures. At this point we need to consider that either:

1. A Type III error has been incurred where the wrong problem was addressed (within a perspective), or that
2. The problem and its nature was never fully understood to begin with.

Failure studies such as the Standish Chaos report (1995) paint a very bleak picture. The original report found that only 16% of projects were success as measured by their adherence to schedule, budget and requirements. Even if this number is low, it is the degree to which targets were not met that should raise questions. For example in 52.7% of the projects, cost over-runs amounted to 187%.

Subsequent studies in 1996, 1998, 2000, 2004, 2006 and 2009 by the Standish group show that project success has increased over the years to just over 30% and that failure and challenged project rates have both decreased. The numbers however still point to significant problems.

Errors of this magnitude and frequency are most likely due to errors in paradigms used in planning, organizing, estimating costs, and scheduling rather than mere control issues. Highsmith (2009) also argues that it is not the projects that are failing as much as it is the project management and planning.

Perhaps some of the most indicative support comes from Jeff Conklin (2005) in his work on Wicked Problems and Social Complexity. His argument for the necessity to gain a better understanding of Wicked Problems and how we deal with them throws the ball directly into the courts of the "tool builders"; those that develop methods and approaches. In his studies of designers working with wicked problems he consistently demonstrated that these individuals did not follow accepted models such as the linear waterfall model of systems design, or for that matter any of the available models. The designers out of necessity, to meet the conditions imposed by the situation followed a near random sequence of data gathering, analysis, solution formulation and implementation.

Ghoshal (2005) argues that it is incorrect perspectives that are at the root of many problems. He argues that common sense has to an extent been usurped by the pretense of knowledge; and that supposed knowledge leads to a wide spectrum of disturbing behavior. This inconsistency of behavior leads to everything from flawed outcomes to downright unethical behavior. Ghoshal and Management (2005) explains that the latter is not only possible, but justified for people undertaking the unethical behavior given the manner in which they perceive a problem or situation.

It is unclear why the lessons that we have learnt over millennia are set aside, paid lip service, or completely forgotten or ignored. Ultimately, if we identify that the behaviors of an organization are emergent, we have to conform to this finding and accept it in its full meaning. We recognize that complete certainty of any problem is not possible; but fail to adequately accept the degree of uncertainty that we generally have to deal with. Many of the problems identified in the failure reports are very likely due to the assumptions of certainty that are made; assumptions that ultimately fail on their encounter with reality. Even after the fact, we seldom question whether it was an irreducible uncertainty that lay at the root of the problems.

This book hopes follow in the footsteps of heretics and challenge the assumptions that underlie our present modi operandi.

1.1.3 Background of the Included Work: The MECS Forum

The materials in this book have a very rich history, and so it would be presumptuous to claim a point in time that the work was started for ourselves.

The work included in this book is derived from the efforts of a group of practitioners and academics that has taken on the scholarly challenge to overcome intellectual inertia and forge new ways to think about and deal with complex situations.

This group of individuals coalesced to form the MECS Forum in 2004. It started out as a small working group at Old Dominion University that had the benefit of being able to meet in the Provost's residence, which had been suitably converted into a "research house". Whether it was due to the beautiful work surroundings and views of the Elizabeth River from the porch of the research house, or the topic that was being discussed, the working group rapidly drew interested people and organizations to itself.

The MECS Forum began as a small think-tank but soon became a community of interest. Since 2005 the MECS Forum has organized multiple symposia, workshops and meetings. The Forum established links to governmental, defense, industry and other societies. It includes active collaborators from diverse universities. The diverse meetings were run with a focus problem at their core. For example, in 2005 a symposium was held with the Joint Forces Command as the sponsoring organization discussing the difficulties of integrating diverse organizations. In 2006 a similar workshop was held in collaboration with the Stevens Institute and CSER focusing on homeland security. In 2008 a workshop was held in collaboration with MITRE at which the topic of interest was "identifying the Paradigm Shift". The symposia, workshops and meetings became the impetus for larger meetings, such as the *MECS Forum Conference* that was held in 2007 in Norfolk Virginia. The MECS Forum Conference is a core source for the contents of this book.

As editors of a synthesized book of this nature it is always a challenge to ensure that a common theme and continuity be maintained; particularly when the diversity of the researchers and writers is taken into consideration. Even though the book is the result of independent work of many individuals and distributed groups, it has benefited from being guided by the similarity of the problems that the different contributors have experienced as well as the collaboration that has taken place through the MECS Forum network.

1.1.4 Intended Audience

This book is intended for scholars, be they academics or practitioners. The book breaches the gap between theory and practice by focusing on:

1. The inherently practical aspects of situations, and
2. The nature of meta-constructs and how the relevance of understanding of the nature of problems on practice.

The manner in which the material is presented makes the book most suitable for researchers, whatever their walk of life. Although everyone on a daily basis encounters wicked problems, the papers and book are intended for individuals with an advanced education. If used as a textbook we would recommend it for upper level master degree or doctoral courses.

The work incorporates many different fields. Wicked problems and complex situations insist on ignoring the well-established academic and disciplinary boundaries that we seem to feel comfortable with. Most problems would typically be said to include a variety of elements, both social and technical. Even in problems that are focused at more specific issues will include a cognitive element that the practitioner, problem solver or research must bring to the table.

1.2 Overview of the Book

1.2.1 Part I: Theory of Complex Situations and Wicked Problems

The first part of the book presents several papers that present theory that underlies complex situations. These theories provide a perspective or interpretative framework for wicked problems. The theories are by no means complete or comprehensive. The study of complex situations and how they pertain to wicked problems found in organizational theory, leadership, management, is an ongoing effort.

The section includes a narrative of complex situations and situations theory that establishes on one hand the boundary of the problem domain that we refer to as complex situations, and simultaneously a proposed theoretical research space or agenda that needed to begin understanding how to deal with situations.

It includes foundational papers that define the interaction of the individual with a problem or entity; forming the situation. This interaction lies at the core situations theory. It establishes the unit of analysis so to speak at a "micro" situations level of study, exploring how our perception of a situation is established, the quality of that representation, and the understanding that it evokes.

The theoretical content presented begins to work at describing how these different perspectives that are formed begin to coalesce into common or shared structures and how the nature of the situation; including both the observer (self) and entity (other than self), influence how we can deal with wicked problems at a "macro" situations level discussion.

1.2.2 Part II: Organization, Management and Engineering in Wicked Problems

Part II of the book focuses on a variety of themes that are recurring within situations and wicked problems. The topics that are presented include: indeterminacy, difficulties in integrating or bridging different perspectives, cognitive and behavior complexities in understanding complex situations, and teleological facets and how we perceive and "interact" with a future state.

As with the theories, this is not an exhaustive. Rather the focus of this section was to explore a select set of concepts in detail to present the nature of the critical thinking that is required when thinking of and dealing with complex situations.

These concepts are presented in papers that discuss practical issues and begin to bridge the chasm between theory and practice. The papers vary including topics related to: sustainability, the complexity of bridging different organizational “levels” in a discussion command at the operational level of war, decision making, leadership, and integration as it pertain to a large military organization.

1.2.3 Part III: Cases and Approaches Related to Wicked Problems

The final section of the book includes work that is case or approach related. It introduced the nature and type of perspective and tools that are required to deal with complex situations.

These papers are intended to highlight the how the tools deal with the incompleteness and uncertainty that are pervasive in complex situations and wicked problems. They also introduce the concept that to deal with wicked problems with any level of effectiveness, it is required that the observer or problem solver think along a spatial dimension, constantly focusing on trying increase the level of understanding of a problem, but also along a temporal dimension to be able to manage the residual uncertainty that is inherent in any complex problem. As such the approaches and tool convey the basic idea of the persistence of such problems.

References

Beer S (2002) What is cybernetics? Kybernetes (MCB UP Ltd) 31(2):209–219

Churchman CW (1967) “Wicked problems”, guest editorial. Management Science 14(4): B141–B142

Conklin J (2005) Dialogue mapping: building shared understanding of wicked problems. New York

Ghoshal S (2005) Bad management theories are destroying good management practices, Acad Manag Learn Educ 4(1):75–91

Hart WD (1996) Dualism. In: Samuel G (ed) A companion to the philosophy of mind. Blackwell, Oxford

Highsmith J (2009) Agile project management: creating innovative products, 2nd edn. Addison-Wesley Professional, Boston

Krell TC (2000) Organizational longevity and technological change. Journal of Organizational Change Management 13(1):8–14

Machiavelli N (1961) The prince (trans: George B). Penguin, London

Plato (370 BC) Parmenides, translated by Benjamin Jowett. Internet Classics Archive. Downloaded from: http://classics.mit.edu/Plato/parmenides.html. July 2011

Rittel H, Webber M (1973) Dilemmas in a general theory of planning, policy sciences, vol 4. Elsevier scientific publishing company, Amsterdam, pp 155–169

Robinson H (2003) Dualism. In: Edward NZ (ed) The Stanford encyclopedia of philosophy. http://plato.stanford.edu/archives/fall2003/entries/dualism/. Fall 2003

Scott WR (2002) Organizations: rational, natural, and open systems, 5th edn. Prentice Hall, Englewood Cliffs

Stacey RD, Griffin D, Shaw P (2000) Complexity and management: fad or radical challenge to systems thinking? Taylor & Francis Group, Routledge

Standish Group (1995) Chaos report. The Standish Group International, Inc. Available at: http://www.projectsmart.co.uk/docs/chaos-report.pdf
Sun T (2005) The art of war by Sun Tzu – Special Edition (trans and annotated: Lionel G). El Paso Norte Press, El Paso
Tolstoy L (1998) War and peace (trans: Louise and Aylmer M). Henry G (ed) Oxford's World Classics, Oxford
Zhang W-B (2002) Theory of complex systems and economic dynamics. Nonlinear Dynamics, Psychology, and Life Sciences 6(2):83–101

Part I
Theory on Wicked Problems and Complex Situations

Chapter 2
A Narrative of [Complex] Situations and Situations Theory

Andres Sousa-Poza

Abstract When faced with intractable problems that are surrounded by high degrees of uncertainty and transience, it is imperative that the nature of the problem be understood. Too frequently, solutions are proposed in the same disciplinary construct in which the problem was framed, and the intractability encountered. Situations Theory is the meta-theoretical study of the conditions that surround a problem. It is not a specific discipline aimed at developing understanding higher levels of understanding of specific phenomena from a well-formulated perspective. Rather it is focused at understanding how such perspectives contribute to dealing with problems encountered in complex situations. The practicality of Situations Theory lies in being able to better judge the effectiveness that an approach will have given a specific set of conditions that might surround a problems, or as a basis on which robust methods and tools can be developed that are congruent with the nature of the problem that is to be addressed.

Keywords Complexity • Complex situations • Wicked problems • Situations theory • Complex adaptive situations methodology • Pragmatic idealism • Paradigm • Worldview

2.1 Introduction

2.1.1 In Breve

In complex situations understanding of the *nature of the problem* is required to ensure that consistent methods and approaches are being used given the conditions encountered, even though an understanding of the problem itself may be limited.

A. Sousa-Poza, Ph.D. (✉)
Batten College of Engineering and Technology, Department of Engineering Management and Systems Engineering, Old Dominion University, Kaufman Hall 242, 23529, Norfolk, VA, USA
e-mail: asouapo@odu.edu

S.F. Kovacic and A. Sousa-Poza (eds.), *Managing and Engineering in Complex Situations*, Topics in Safety, Risk, Reliability and Quality 21, DOI 10.1007/978-94-007-5515-4_2,

For this narrative, situation refers to the condition(s) that are encountered by the participant-observer. Situations theory refers to the set of theories, models, tools, and know-how that have evolved from research of wicked problems that provide an alternative perspective of such problems. Situations Theory is the study of the *nature of problems*, from a meta-theoretical point of view. As such, it is not a discipline aimed at solving a problem or set of problems, but rather a meta-discipline studying the efficacy of the understanding generated by other disciplines. That is not to say that there is limited practical value since this understanding of the *nature of problems* serves to inform a rich methodology; namely, the Complex Adaptive Situations Methodology (CASM). It has also enabled the conceptual and prototypical development of tools, methods and approaches for a variety of problems including organizational structuring, decision-making, risk management, operational integration, situational awareness, and other fields.

There is a highly practical and pragmatic focus to Situations Theory that generally transcends purely disciplinary studies. The interest lies in maximizing the possible comprehension and understanding that we have of a problem, irrespective of the discipline, sense-making, worldview or otherwise labeled process that served to form the representation of the problem. Interest in disciplinary improvement becomes subservient to this practical imperative, however, for a simple problem to exist, it can be assumed that an adequate representation has been formed to be able to solve the problem.

2.1.2 Background of the Problem

It is not uncommon to start research on a problem, and while working towards a goal, end up heading in a completely different direction. Ironically, the very non-ergodic nature of research not only underlies the indeterminacy that results in these changes in direction, but also turns out to be one of the many attributes of the problem that was being studied.

The work that lead to the research of complex situations and the formulation of situations theory, started out quite innocuously as an advanced study of interoperability of systems, or system of systems engineering (SOSE). The intent of system of system engineering was to develop design based approaches that would enable the engineered acquisition of capabilities through the integration of existing and new technologies, both in the form of hardware and software, and organizations. Acquisition of capabilities in this manner has and remains problematic due to the intricacies that are generated when integrating disparate components. Difficulties arise from a variety of factors, for example:

- The integration of technologies of different generations. Older technologies may overly restrict newer technologies. Reverse compatibility may not be possible.
- There may be differences in the synchronicity of problems forcing changes to underlying processes.

- Conceptual difference in data may make sharing of information extremely difficult if not impossible. Different frames of reference may defy mapping to each other.
- Integration of independent social organizations may encounter a variety of socio-psychological barriers ranging from an unwillingness to share by stakeholders of the different entities, to politics and policies that restrict integration or even interaction through normatively or regularity means.

Keating et al. (2003) established a baseline for system of systems engineering, identifying the different problems, and perspectives for the field. Unfortunately little consensus could be generated between researcher, research sponsors, and even practitioners that were seeking means to propagate any capabilities. Different schools of thinking about the problem rapidly emerged and have not seen significant convergence in methods (Sousa-Poza, Kovacic, Keating 2008). To confound matters further, a singular concrete definition of what constitutes a system of systems has failed to emerge.

Two important reports that strove to define SOSE were generated. The United States Air Force Science Advisory Board (AF SAB) published a report on SOSE in 2005. To ensure consistency in the methodology, it was required that problems be "predictable", and be approached in a "predictable manner". This approach eliminated many systems of systems based capabilities acquisitions that face extensive emergence and/or self-organization. A second report published by the Trans-Atlantic Research and Education Agenda in Systems of Systems (T-AREA-SoS) (2012) titled "State of the Art on System of Systems Management and Engineering" takes a much more inclusive perspective on integrating diverse entities, identifying many problems and areas that require further research.

Case studies and implementation efforts resulted in the conclusion that *engineered* integration of technologies and organizations might only be possible for a subset of system of systems type problems. Most problems that were being addressed for defense and security type problems repeatedly encounter serious problems, for example:

- The Integrated Deepwater System for the US Coast Guard,
- The SBInet for US Department of Homeland Security, and
- The UK FiReControl project.

2.1.3 *The Problem*

It was becoming apparent that the development of SOSE was facing attributes that it might not overcome (Sousa-Poza et al. 2004a, b; Sousa-Poza and Keating 2004; Keating and Sousa-Poza 2005; Kovacic, Sousa-Poza, Keating 2007a). Whether it is due to technological, semantic, social or psychological factors, the attempt to engineer complex and numerous entities into an integrated capability delivering organization was turning into a Helleresque catch-22. To integrate and ensure the interoperability of increasingly diverse, heterogeneous components, increasingly intricate and involved designs would be required to the extent that the "complexification" would ultimately become a liability.

This liability is exemplified in Mr. Ray Ozzie's discussion on the future of Microsoft Operating Systems (http://ozzie.net/docs/dawn-of-a-new-day/), and consequences of the intricacy being incorporated. To ensure that all the different OS components and application function together, and that they are reverse compatible requires a great deal of "complexification". He indicates that these designs are resulting in extremely *fragile* solutions instead of realizing ever-greater agility, flexibility and robustness. In other words, there appears to be a limit of our comprehension of integration and interoperability type problems and solutions.

Our research arrived at the conclusion that a different line of questioning was going to be required, namely, "what are the attributes of the problem that (1) impose limitations on the methods being used, and (2) given a set of attributes can new methods be developed that circumvent the limitations?"

Efforts to understand the nature of the problem better, and to identify solution approaches led to the recognition of facets of engineering and management other than interoperability facing similar complexity and uncertainty. For example, Conklin (2006) found that in wicked problems, as Rittel and Weber would have prognositcated, work did not flow in a nice linear manner according to the waterfall model. In fact, it was found that problem solvers nearly concurrently maintain facets of analysis, problem formulation, solution development and implementation. This finding is very much in line with the description of the attributes of wicked problems that Rittel and Weber provided. For example:

1. In an environment where a there is *no definitive problem definition*, due to whatever reason, the formulation of the problem becomes an ongoing elaboration of the awareness and understanding and it would be expected that the problem solvers constantly switch back to, or even maintain a problem identification phase.
2. The problem definintion and the solution become synnonymous and the *choice of an explanation determines the nature of the problem's resolution.*
3. The *lack of a stopping rule* means that there is no means to define the transition between states, particularly since no clear outcome state might be definable.

Attempts to fomulate more effective approaches to deal with wicked problems was first directed at improving existing based on the rudimentary understanding that had been developed of wicked problems. For example, to overcome the problem of having many complementary perspectives, a Unifying Framework approach was developed. The rationale behind the Unifying Framework followed the lead of extensive seminal work in diverse fields, such as General Systems Thoery (Skyttner 1999; Von Bertanlanffy 1976; Checkland 2006). This attempt resulted in the theoretical ability to combine two or more perspectives. The investigation, however, led to the conclusion that the integration of perspectives is extremely inefficient. In extreme cases, the loss of fidelity imposed on the individual perspectives might be enough to make them useless for the function that they are to support, while not guaranteeing that a sufficiently strong contribution would be made to the common perspective to justify any loss (Kovacic, Sousa-Poza, Keating 2006, 2007a, 2007b). The lesson learnt from these efforts was that an attribute of a complex situation is intrinsic, and it is not possible to overcome the attribute by an increased effort using the method that led to the recognition of the attribute in the first place.

It also became apparent, that as new approaches were formulated to deal with one facet of a problem, many others would be affected. For example, to maintain integrity in organizational structuring and reduce the amount of conflict, it is often suggested that self-organization be embraced using policies and norms to regulated behavior within the organization. For this to be possible, changes are also required to planning, resource allocations, human resouces, and decision making.

The problem, in brief, stems from the limits to the understanding that any perspective is able to generate, which limits what can be done to address a problem encountered within the situation. The problem is generated when we fail to recognize the limitation, or if we fail to adjust our paradigms and behaviors in spite of having identified them.

2.1.4 *Need for a New Approach*

All lines of reasoning resulted in having to rethink "everything". For example:

- Imposing a new paradigm on how to structure and organize to better deal with complexities encountered in interoperability allowing for greater self determination and self-organization,
- New methods would be required for decision making to allow for the inherent uncertainty that would be generated by the innumerable alternatives that would be generated,
- And, new planning and goal attainment methods would be needed to deal with the lack of a clearly posited end-state or the lack of an identifiable transition.

The dominant principle became that new methods would be required that were consistent with the problem, and in turn with each other.

The importance of the evolution of a paradigm that has a comprehensive impact is also underscored by the recognition that in a practical setting, it is not possible to isolate functions. The decision making approach cannot be altered without taking into consideration the impact on planning, control, etc.

2.1.5 *Overview of the Paper*

The purpose of this paper is to provide an overview of complex situations and situations theory. The focus is on:

1. Describing
    (a) Situations
    (b) Situations Theory

2. Purpose of situations theory
3. Highlighting some of the salient features of situations theory
4. Methodological considerations in evolving situations theory
5. Research agenda of situations theory
    (a) Micro-Situations Theory
    (b) Macro-Situations Theory
6. Consequence of ST: Call to Heretics

2.2 What Are [Complex] Situations?

A situation is more readily described than defined. When dealing with topics that scratch the edge of what may be comprehensible, it is inevitable that the justification of any definition, like an Ouroboros, will ultimately turn on itself and become part of its own justification.

To ovoid this 'complication', it often becomes necessary to bound and constrain a problem or context to ensure the consistency of the definition, whilst hopefully not forgetting that such a bounding will cause an obvious incompleteness, a separation of the definition derived from the bounded space from that which we initially intended to define. The definition will ultimately only be as good as the bounding that we impose. In those cases where any bound will detract from that which we originally set out to define, the definition will suffer and fail to fully convey that which we intended to state.

Abstractly, a situation is defined as a set of conditions that we expand on with the requirement that an individual 'is' or 'becomes' cognizant of the set of conditions. It is consequently an experiential construct that requires the participant to simultaneously establish an awareness of that which is being experienced and in so doing a self-awareness that bring the experience into being. It requires simultaneously having a perception be formed through the observation of that which the individual is experiencing as well as the [cognitive] imposition of the individual on that experience.

On a metaphysical or microcosmic scope, situations are a transient and near persistent state that are manifested when an individual becomes aware of a set of conditions, or when that individual imposes themselves to form a set of conditions. They are the combination of an unbounded participation of a person in life (noumenon) as well as the cognitive and bounded observation of that life (phenomenon).

The situation, as best we can, quite simply establishes "*that which is*". We refer to this as the *Reality Is* principle. Due to its largely unbounded nature, we recognize that "*that which is*" is in general, and in most cases, invariably much greater than that which we will be able to form an understanding of (Sousa-Poza and Correa 2005; Brewer 2010). In very select, special, and constrained cases do the conditions manifest themselves in such a manner that the understanding of "*that which is*", is able to form a perspective that is representative of "*that which is*". This would be referred to as a *Simple Situation*. This also implies that the complex situation becomes the general case, making the word complex somewhat redundant and unnecessary. The distinction between a simple and [complex] is thus defined by the degree to which comprehensibility and understanding of the situation can be established.

2.2.1 On Situations and Wicked Problems

Complex situations, such as those that Rittel and Weber describe in their work on wicked problems, refer to problems of a highly practical nature. Not that they are not of interest to researchers, academics and other scholars, but rather that there are situations in which the bounding of a problems requires the imposition of assumptions and simplifications that may very well reduce its practical value; even if in so doing make them more acceptable from an academic perspective. To meet the criteria of knowledge researchers are compelled to carefully govern the context in which the research is conducted to ensure the consistency and reliability of their conclusions.

Wicked problems, as have been described in literature, have an intransient and a perplexingly incomplete and intractable nature to them that make their study extremely difficult. It is possible to extract elements of these problems and isolate them from everything else that is happening to possibly study some facet of what might be transpiring. This separation is at once necessary, but also dangerous in the sense that the conclusions that are derived within the bounded space may be barely relevant to the original problem being studied.

Studying situations fundamentally becomes similar to the development of our understanding of wicked problems (Kovacic, Sousa-Poza, Keating 2008). The concrete, intransient statements that we can make reflect the nature of the problem. These focus on:

1. The *wickedness* of wicked problems
2. What makes wicked problems, '*wicked*', and ultimately,
3. How a person should proceed given the *wickedness* of a problem

Wicked Problems are at once a form of situation, just as situations come into existence through the awareness of the existence of a wicked problem.

2.2.2 On Situations and Systems

The terms "situation" and "system" are semantically quite distinct. A system can be referred to as a set of inter-related components that together generate a behavior. From this a variety of different perspectives or interpretations are possible.

Commonly a strong empiricist bent is present where the system and that which it is supposed to represent become synonymous. Another interpretation would be much more rational in nature, where the word system is merely a representation. From an analytic point of view the system is representational of an observed set of components and their ensuing behavior. Alternatively, however, the system may precede any observable entity, as would be the case in a design.

In all cases, a system is ultimately a bounded representation for which there is the "system" and the "environment". The frequent inclusion of the environment in systems models actually forces the inclusion of the environment into the system or a layered view that would include the system, the environment, "and everything else that could not be considered". Systems, as a discipline, is a means by which we attempt to make sense of something. The simplifications and assumptions made in the bounding process and the ensuing focus on the bounded system ultimately lead to the risk of turning this into an academic exercise, studying (often in great detail) a system that no longer has any bearing on actual events.

Conversly, a situation differs from a system in that it is less about the workings within a perspective (or this case system), and much more about the manner that the bounded perspective interacts with the set of conditions, in other words the situation, from which it is derived. Assuming that a system representation is intrinsically as good as it can be, the question that arises is how much comprehension of a situation is possible. Accepting that there may be limitations or flaws in the representation itself, we also recognize that the degree of understanding may not fully reflect what might be comprehended (Brewer 2010).

2.2.3 Research of Situations: A Segue to Situations Theory

The consequence of this has been that in general, research on complex situations has studied the meta-constructs of complex situations; in other words, *the nature of situations.*

Both the focus on studying the meta-construct and the sheer intractability of many of the topics being addressed helped evolve a unique perspective on how research is undertaken. Methodologically a reliance on rational coherence is necessary to ensure that the inferences being drawn are consistent (Sousa-Poza and Kovacic 2008).

The difficulty of maintaining a continuous tie to actual or practical conditions also presented challenges. Research within a laboratory or other bounded domains is possible. Maintaining the imperative of practicality and where necessary the participatory construct as a necessary part of a situation does present challenges within social settings. Elaboration of quasi-experimental techniques, as well as specialized laboratory environment became imperative to ensure that even when a virtual environment is necessary, that a moderating assessment of the impact of bounding the problem space can be effectively provided, through for example, the use of individuals familiar with a problem (Sousa-Poza and Kovacic 2008).

2.3 An Overview of Situations Theory

2.3.1 Purpose of Situations Theory

The purpose of Situations Theory is to provide a set of theories and models to facilitate the understanding of the nature of a situation with the practical intent of improving practice in [complex] situations. The guiding principle of Situations Theory is to maintain within the developed constructs the natural tie to reality that (1) is necessary to understand a dynamic, transient, context specific, situation, and (2) an individual, as a participant, will continuously experience. Situations Theory is intended to inform with respect to a reality in which the individual participates, not solely a construct of the observer.

The importance of this principle is driven by:

1. The incorporation of the participant and maintenance of the observer in the problems (Kovacic, Sousa-Poza, Keating 2007, 2008)
2. The practical nature of the problems
3. The paradigm captured in Pragmatic Idealism, that led to the present state of understanding of Situations Theory and Situations Theory, helped to influence (Sousa-Poza and Correa 2005).

Under certain circumstances, where the participatory influence cannot be externalized (Nonaka et al. 2001), the situation and its nature will only be understood when maintained in situ to retain continuity of the context.

Situations Theory provides principles, models and means to interpret the nature of the situation (see for example Brewer and Gheorghe 2011).

2.3.2 Representations of Reality – RDP Model

Important in the evolution of situations theory were the ideas presented in Sousa-Poza and Correa (2005). The basic ontological construct stipulates that at the point that awareness and self-awareness occur, a domain, D, is established that encompasses the comprehensibility, C, that a perspective *potentially* could provide of the situation (Fig. 2.1).

The extent that the domain captures comprehensibility of a problem is contingent on the nature of the bounding. The assumptions, simplifications and other strategies that come into play when the domain is formed will influence the approximation distance, A′(d), between the domain and reality. A′(d) can be interpreted as a proximity measure affecting the resultant comprehension.

The degree to which the perspective can capture the comprehensibility offered by the domain will influence what is ultimately understood, U, of the problem. If the approximation distance, A″(d) between the domain and perspective is large the understanding of the problem will ultimately also be significant.

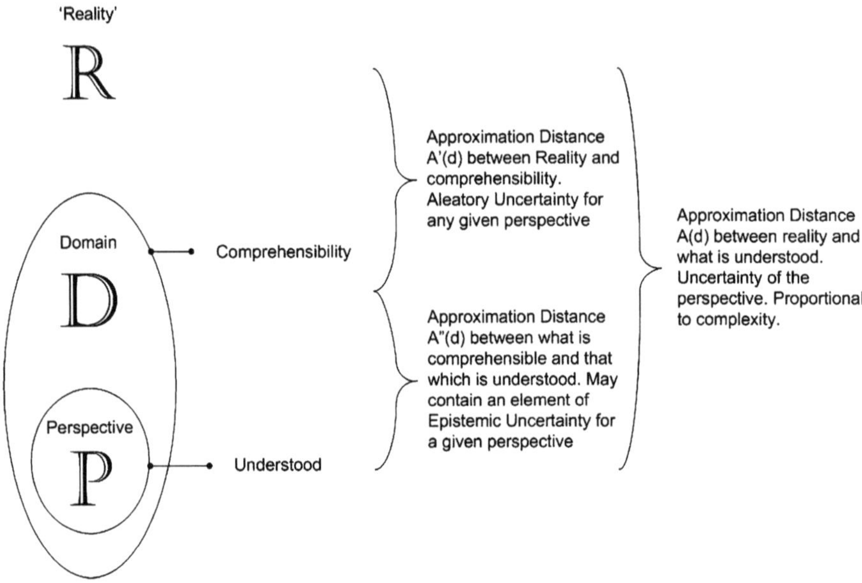

Fig. 2.1 RDP model

In situations theory the total Approximation Distance, A(d), that is formed by the confluence of A′(d) and A″(d) is of interest. Minimizing A(d) is one of the primary objective of situations theory. Understanding the interwoven relationship between the two approximation distances, A′(d) and A″(d) is crucial to understanding how A(d) may be minimized. Since the generative process that forms the domain and perspective applied to both, D and P cannot be treated as independent constructs. Consequently focusing on only A′(d) or A″(d) may have a detrimental effect on the overall understanding of the problem.

For example, a highly structured, tight bounding might be sought for D to ensure that it can be well understood. This would minimizing A″(d). This domain formulation may however require extensive assumptions and simplifications to be made that result in a large A′(d) and consequently the domain might be well understood but not the problem.

Similarly, a much more encompassing domain might be established to better capture the problem, minimizing A′(d). The inclusion of great variability, heterogeneity, etc. in the domain might however result in very little of it being understood. In this case there is a high *potential* comprehensibility but low understanding.

Ultimately it is imperative that the individual recognize the forces at play and come to understand what strategies minimize A(d) for diverse conditions and problems.

2.3.3 *Sense-Making*

The formation of a perspective and understanding of a problem takes place through a generative process. It incorporates the philosophical, sociological, psychological, physiological and other factors and predispositions that form the individual's cognitive foundation. Brewer (2010) refers to the perspective as the Cognitive Representation of Reality, and defines cognitive in an extremely expansive manner reminiscent of ancient Greek philosophers' definition of psychology.

Once a domain and sense making, generative process is established, the best outcome for the objective of minimizing A(d) will be A′(d) (assuming that increasing understanding by reconsidering P will not affect the comprehensibility captured by the domain). If it is established that the residual [aleatory] uncertainty is too high, a change in the domain is required. This requires a change in the generative process that will ultimately alter D and the situation. Whether or not the original problem is still perceived will be contingent on the outcome of the shift in paradigm and resultant perspective. In other words, a greater understanding might be obtained at the cost of "loosing" the problem. Obviously further refinement of the generative process and ensuing Domain-Perspective dyad would be required.

2.3.4 *The Participant-Observer Dyad*

This process of bounding or making sense of Reality also results in the individual adopting two roles: that of an observer, and that of a participant (Fig. 2.2). This relationship between participant and observer is central to situations theory. It can be defined and described as a syncretic construct. The individual is defined by both roles, each role making up a side of a coin. The individual and their perspective cannot fully be understood by just one of the facets. Recognizing both is however not a guarantee depending on the degree to which the perspectives formed by the roles are complementary. The necessity of maintaining both roles is predicated by the extent that the two roles are reconcilable, or whether the role of the participant can be externalized into a more explicit element of the observer's perspective.

The generative process is a means by which sense can be made of all of the inputs and influences acting on the individual both in their role as an observer and as a participant. It is not solely a formal, explicit process; relying also on experience, intuition, expertise as well as formal reasoning to organize reality into a perspective.

In problems that are highly context sensitive, it becomes imperative that the individual undertaking the problem is also a participant. This is particularly true if an observer alone cannot codify the context of participation.

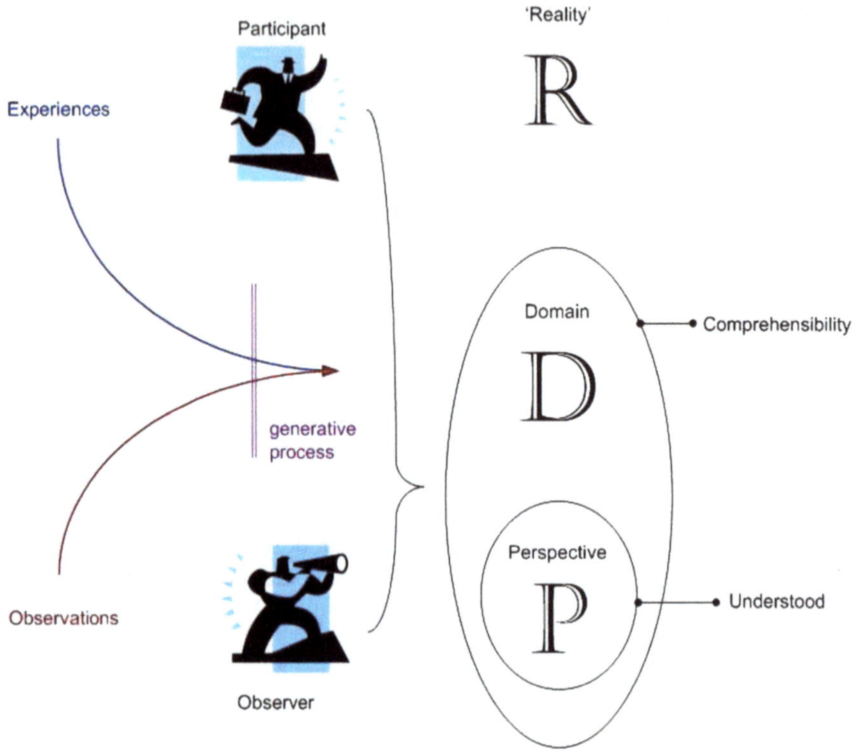

Fig. 2.2 The participant-observer

2.3.5 The Syncretic Imperative: Paradoxical Coexistence

Paradox can exist in two forms:

1. The first is when understanding of a problem and expected behaviors are found to be incongruent with resulting "Real" or observed behavior (Fig. 2.3). Typically an error in the perspective would be assumed. However, since the expected behaviors are predicated by the understanding derived of the same generative process that defined that which is comprehensible within a domain, the question of right and wrong has a secondary layer. An expected behavior can be found to right based on the domain and comprehension from which it is derived. In other words is it consistent within the construct of the domain. This still allows for a disparity with observed behaviors given the approximation distance between the Domain and Reality. The expected behavior is both incorrect and correct.
2. The inclusion of several syncretic structures results in the potential for conflict between perspectives that might also be paradoxical. The condition is possible where two perspectives result in two irreconcilable descriptions of an entity, or

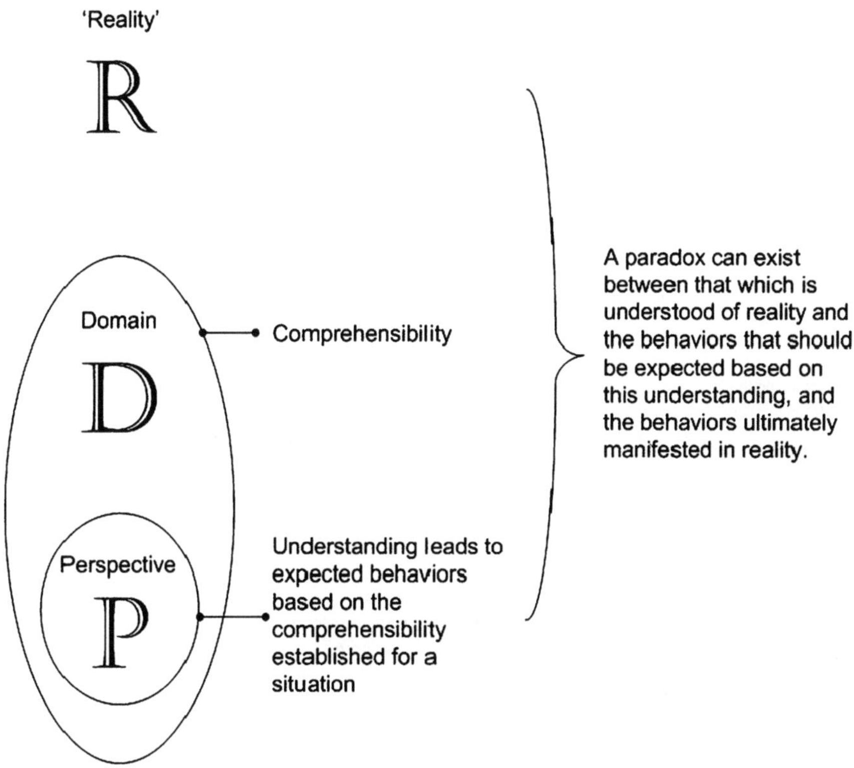

Fig. 2.3 Paradox of counterintuitive conclusions

two expected behaviors that cannot coexist (Fig. 2.4). The expected behaviors may be fully consistent within their domains, making them both correct. This dilemma is at the heart of many complementarities and dualities (Bohr 1949) in which two behaviors should not be attributable to the same entity. For example, an entity should not be able to behave as a wave AND as a particle, since particles do not exhibit wave like behavior, and waves do not exhibit particulate behavior. Within their respective domains, the wave and particle behaviors may however be completely consistent with expected behaviors and thus judged to be correct.

These dilemmas highlight an interesting facet of complex situations that also remains consistent with the RDP and other models, namely that problems can be correct, yet wrong, or even luckily, incorrect and right!

- The terms correct and incorrect are applied when judging whether a description or expected behavior is consistent with the conditions and comprehension established within the domain.
- The terms right and wrong are applied when judging whether an expected behavior or description is consistent with observed behaviors or conditions.

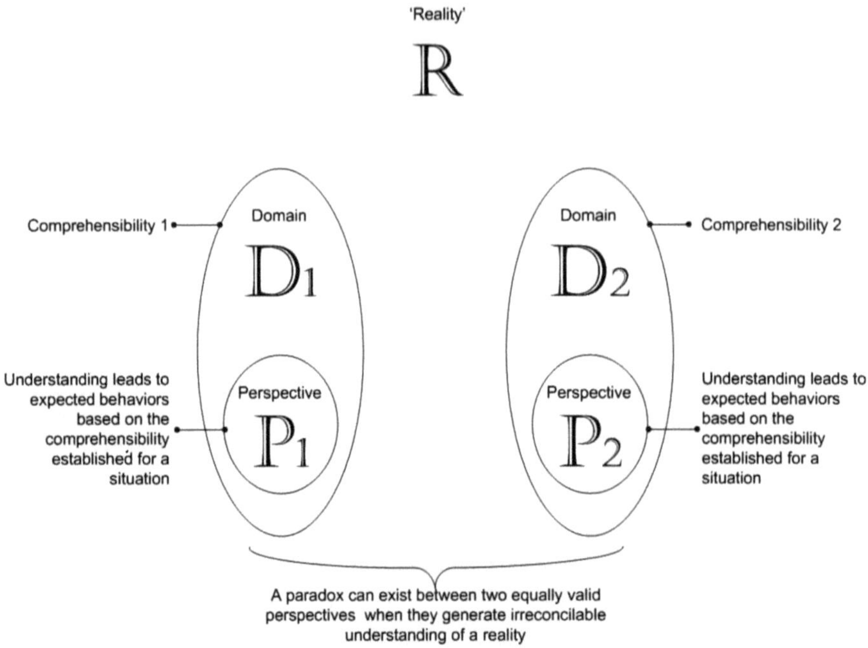

Fig. 2.4 Paradox of irreconcilable perspectives

A further distinction can be made which refers to the approach or methods applied, where the judgment is on the congruency of the approach to the conditions that surround a situation. A method that is not capable of dealing with the conditions of a situation is referred to as incongruent or misaligned.

2.3.6 Situations Theory View of Knowledge, Understanding and Situations

Situations Theory uses understanding as the core, practical output that is derived from the activities of observing, studying, researching, participating, and acting by an individual in a situation (Sousa-Poza and Correa 2005).

Participation and Observation take place within the Noumenon (from Kant as per Oizerman 1981) and Phenomenon elements of the situation respectively.

Observation is a conscious, explicit activity. Observation might include any effort to gather explicit information and data. Participation is an activity that takes place at a subconscious level and is inherently implicit. Participation includes the experience derived from working and living within a situation.

At the simplest level, the individual acquires and is part of the inputs that must be made sense of. Although the diagram (Fig. 2.5) infers that there is a sequential

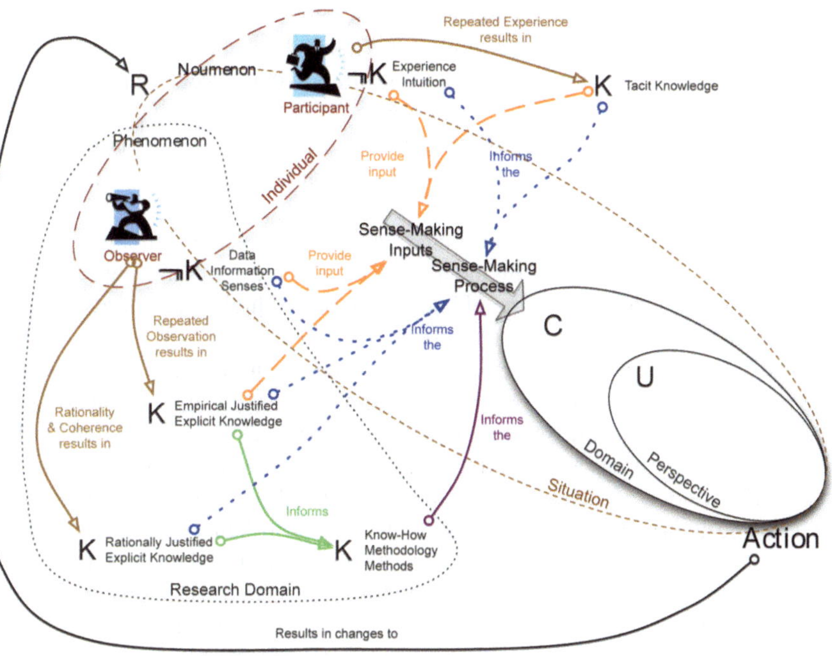

Fig. 2.5 Dual nature of understanding

step in identifying the situation, followed by the formulation of the sense making approaches and generative process, the generative process is part of the mutually generative elements that come to be at the inception of the situation.

Through the generative process/sense making activities an understanding is formulated of the situation. It is this understanding that forms the basis for practical action (Sousa-Poza and Correa 2005; Padilla 2010, Padilla, Diallo, Sousa-Poza 2011). The understanding incorporates, participatory and observed elements. The sources of input may include items that are knowledge (*K*) and items that are categorically not-knowledge (¬*K*). Knowledge in this model is defined as an item that can be considered to be a Justified True Belief (Brewer and Sousa-Poza 2009). Where knowledge is available, assuming that sufficient evidence exists that the knowledge is appropriate for the situation in question, one could assume that a high utility should be derived from the knowledge (Brewer and Sousa-Poza 2009).

The contextual sensitivity of situations, where the individual's predispositions play a significant role, makes the use of all but highly generalized knowledge difficult. Transferability of knowledge may be difficult where sufficient recognition of important contextual factors are missing (Brewer and Sousa-Poza 2009).

Since individuals maintain a participatory role it can be expected that the continued experiences that an individual has will have just as significant effect on the situation as that which is learnt is assimilated in the perspective and the sense making processes. It can be expected that situations, other than in some very specific, highly static cases, will be highly transient. This high level of transience will make the utilization of knowledge difficult if not impossible (Brewer and Sousa-Poza 2009).

Consequently, from a practical point of view, the use of a greater variety of inputs will be required. The burden will ultimately fall on the generative or sense making processes.

Knowledge acquisition in such environment is possible although it will be affected and possibly limited by the nature of the context and transience. In a participatory role the dominant form of knowledge will be highly tacit derived from the indoctrination and internalization of effective behaviors.

The observer in their interaction with the phenomenological element becomes the dominant source of knowledge. Although the transience and contextual sensitivity may impose the requirement that knowledge be established of large number of observations and will likely be extremely abstract in nature. Empirically based knowledge is derived from the repeated observation of behaviors and results and is justified through correspondence (Millikan 1986; Padilla, Diallo, Sousa-Poza 2011). Rationally based knowledge is informed through the interaction of an individual with complex situations (in general), and other sources. Rationally based knowledge depends on coherence to provide justification (Dauer 1974; Sousa-Poza et al. 2008a, b). A further form of knowledge is formed that is methodological or pragmatic in nature. This knowledge focuses on "how" and will dominantly influence the sense-making processes rather than inform the perspective itself.

Important is the differentiation of knowledge and understanding in complex situations. The two have distinct semantic meaning with understanding (if it exists) providing the causal inference required to be able to make decisions and take other actions (Padilla 2010; Padilla, Diallo, Sousa-Poza 2011). Knowledge on the other hand plays a role in supporting understanding, although its utility will vary based on the situation. Even though knowledge may be acquired from any situation, there is no guarantee that it will provide practical utility at a later point in time, or for other problems.

Confusing not-Knowledge with Knowledge, or failing to determine whether an item is knowledge or not-knowledge, from either a participatory or observer role may compound ineffective or incorrect conclusions and behaviors since the treatment of knowledge and not-knowledge is and should be different. Ultimately, a higher degree of trust should be possible with knowledge than not-knowledge based on the justification required in its establishment. This trust must affect the manner in which knowledge can be used in decisions and other actions. The scarcity of knowledge in highly transient, context specific situations should, however, also be noted. Consequently, it is important to recognize the criticality of being able to formulate understanding from not-knowledge. Given that not-knowledge lacks the justification that would make it knowledge means that it should be treated differently, possibly with less trust, but it must be dealt with none-the-less.

Table 2.1 AALA model

Reality is: \ Asserts problem is	Not understood	Understood
Comprehensible	High complexity Epistemic uncertainty Justifiable action is to learn	Low complexity Low uncertainty Justifiable action is to proceed within what is correct in the domain
Incomprehensible	High complexity Aleatory and epistemic uncertainty Justifiable action is to adapt the perspective through the adoption of a new paradigm/generative process	High complexity Aleatory Uncertainty that may not be recognized by the perspective Justifiable action is to assess the perspective and domain

The level of comprehensibility and understanding will justify a variety of actions (Brewer 2010) (see Table 2.1: AALA Model).

In extreme cases it may be difficult to make an accurate determination of the degree of comprehensibility of a situation, or the degree to which an individual believes that they understand a problem (Brewer 2010).

Taking action of some sort will often be the only way to answer this question and provide for a way forward. There are obvious risks involved in taking marginally informed decisions to be able to study their consequence with the intent of resolving uncertainty. Again the reliance on a "live" context, of learning and making decisions in situ make studying the problem in "safe" environment difficult, although advanced in decision environments that account for some of the contextual sensitivity and transience are conceptually possible.

2.3.7 The Situational Construct Model

There are two major ontological elements that contribute to our understanding of situations. The first is the RDP model discussed earlier in this paper, and the second is the situational construct model (Fig. 2.6) that elaborates on the interaction of the observer and the participant with a teleological element (the solution form) (Sousa-Poza 2007). The elements are often difficult to reconcile and quite possibly complementary.

The situational construct relies on an assessment of alignment (Padilla et al. 2007) between the three elements to form an a priori determination of the degree of complexity of a problem by assessing the potential sources of tension as misalignments between three components: the entity (in which the individual participates),

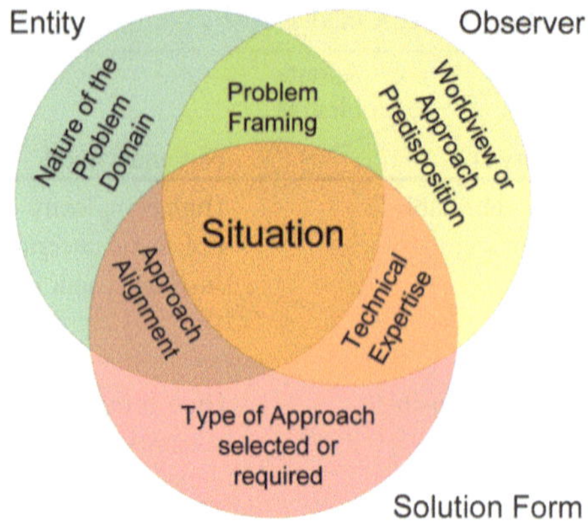

Fig. 2.6 Situational construct

the observer (representing the role of the individual separate from the entity), and the solution form (providing an impetus for the formation of the situation).

The situational construct also highlights that the source of complexity in situations theory is derived from studying the confluence of at least two interacting elements. It is in the alignment of factors, for example, expectations with what is possible, where complexity is generated (Padilla 2007).

As such, an entity, be it an artifact such as a pencil, a machine such as an aircraft, a sociotechnical platform such as an aircraft carrier, an organization such as a country, cannot be deemed to be simple or complex until it is made clear what the expectation is for a solution form and how an individual as a participant and observer combine with this as a situation. The pencil may turn out to be more complex than an aircraft based on what is required.

The three elements of the situational construct include:

- The Entity: which is defined by a set of meta-attributes of the "participatory" element of the problem.
- The Observer: is characterized by the individuals worldview, predispositions, capabilities
- The Solution Form: which as an element includes a teleological components for the situation, establishing the purpose of the situations existence.

The three elements are mutually generative and as such, influence each other significantly in many cases, but also leave the potential for significant misalignment when, for example, one of the elements is subject to external definition.

The first order interaction between the elements provides for a basic assessment of alignment.

- The problem framing is the result of comparing the attributes of the entity with those of the observer. A variety of comparisons are possible ranging from capability (observer is a mechanic, and the entity entails a socio-psychological problem) to worldviews (observer is an extremely detailed, analytic thinker, and the entity is irreducible).

- Technical expertise relates the requirements imposed by the solution form both as a teleological element and based on the attributes of the mean that will be required to reach an expected solution. This may include comparisons of skills, worldviews, predispositions and other factors.
- The Approach Alignment compares the attributes with the attributes of the means selected for the solution form. The comparison may for example be between the requirements of the mean/method to clearly define an end state, which fails to align with the entity that might be non-ergodic.

Ultimately, the complexity and degree to which a problem is understood requires an assessment of all of the elements and their alignment in either a second order analysis, or a qualitative comparison of the combination of all three elements.

2.3.8 The Environment

We understand that, in general, a situation is a highly transient, contextually sensitive construct that is tightly enmeshed with an individual who acts both as an observer and a participant. Individuals that do not participate in the same problem or "reality" space will be limited in their comprehension and understanding to what can be extracted from extrinsic, observable items, and what is lost through the lack of participatory experience. Furthermore, even individuals that do not share the same experiences or are predisposed to think differently will not perceive the situation in the same manner.

Even in the case where there is only one individual, because of learning, changes in the reality, and changes in predisposition, we cannot assume that the perspective at one point in time can be shared with that at a different point in time. Similarly, different generative processes may be required to make sense of the past and present (possibly and empirically driven worldview), and the future (possibly a rationally driven worldview). The different perspectives that are formed may be extremely heterogeneous.

There are two important directions that this concept of the shared understanding leads to. The first is one in which an environment is established to deal with many component perspectives that exist over a period of time; past, and infinitesimally short present and a future. The second relates to a spatial construct that combine many different perspectives at some point in time, where different individuals might hold those perspective.

2.3.9 Fundamental Constructs of the Environment

The relationship between the Environment, which forms a Domain-Perspective (DP) dyad and Reality (R) is of a Syncretic nature establishing R-DP. This formulation implies that R-DP has a singular existence where R exists in DP inasmuch as DP exists in R. It also highlights that R and DP are not [generally] the same and at any point, R may be manifested independently of DP (as a form of participative transcendentalism), as DP may manifest itself independent of R (as a form of observational separatism). From first principles, however, it is co-existence of R-DP that is dictated by Pragmatic Idealism (Sousa-Poza and Correa 2005) (1. non-extremism principle; 2. Awareness principle), rejecting the extremist position that either of the elements may fundamentally be attained separated from the other. In the specific or simple condition we recognize that the differences or differentiability between R and DP may [tend to] disappear, at which point, an orthodox stance is acceptable; but is arguably only so given that R can be [quasi-] fully subsumed into DP, or vice versa, and consequently even the orthodox condition is arguably a syncretic condition although it need not be recognized as such.

The first implication of this is that there is a difference between "that which we say" and "that which we do" (and vice versa) for which a balance must be maintained. This difference establishes a saddle point between "free will" and "fate or the preponderance of inevitability", normative behavior and naturalist behavior, action and consequence, intelligent design and evolution, and so on. For the syncretic construct to maintain its existence, it is therefore necessary that we accept that environments can be established, but must accept the influence of being enacted (participation exists within observation), or that environments can evolve, but must accept the influence of being recognized (observation exists within participation).

The second implication of this is that an environment, as a fully evolved construct, requires R and cannot fully manifest itself in cross-sectional conditions (point in time; discrete; observation), but requires its enactment longitudinally (over time; continuity; participation). In the specific simple condition where the observational separatist construct is a sufficient manifestation of Reality we can still fully argue the existence of R as a [quasi-] fully subsumed construct of the DP. This is the condition on which typical analysis based methods rely in which the DP is relied upon to project future discrete DP. Similarly in the specific simple condition where the participative transcendental construct is a sufficient manifestation of DP we can argue the near complete incorporation of the DP construct into R. This is a condition in which we trust our actions to unintentional, subconscious behavior, for example walking, running, swimming, or as is often the case for situations in which rapid response is required such a life and death situations encountered by soldiers. For such situations, highly conditioned responses are relied upon that can be developed through extensive exercise and repetition until a behavior becomes 'second-nature'.

In the Environment several potential boundaries exist. From an empiricist point of view by the set of all perspectives relevant to the situation, and from a rational perspective by the bounding that encompasses all relevant [manifesting] perspectives. For these two to coexist, a further boundary must be present within which both perspectives are present. This boundary is established by conceptual bounding and at this point we assert that is based on the/a value premise (). At minimum, an environment will contain a perspective, a complement to the perspective, a synthetic perspective formed from the complement pair, and a synoptic perspective that encompasses the complement pair.

[Note: Each perspective is in its own right an environment, since a perspective can form or be defined by its own synthesis or synopsis. The de facto evolution of such an environment would, however, indicate an extraordinary simple condition in which no distinct complementary perspective evolves. The establishment of such an environment would either have to assume a similarly simple condition in which any complementary aspect to the perspective is subsumed in the perspective, or a condition of imposed orthodoxy, where complementary views are fundamentally ignored.]

2.3.10 The Environment – Temporal Integrity

The relationship between R and DP, and the relationship that we have formed between the rationally and empirically derived manifestations provide a means to describe (1) the progression of perspectives and activity over time, and (2) a teleological differentiation based on the dominance in the formation of perspectives (see Stacey et al. 2000 for an extensive discussion on Teleology).

Within an environment (DP) we have formed distinct yet interdependent elements based on local, synthetic and synoptic perspectives. The differences in worldview that drive the formation of the synoptic and local perspectives will result in irreconcilable, yet related views. The synoptic perspective captures the set of possibilities and is representative of the Realmoglichkeiten (Hartman 1949) (Real Possibilities), whereas the set of local perspectives is representative of the Realwirklichkeiten (Hartman 1949) (Real Actuality). In the general case of a [complex] situation, there is interplay between possibilities and outcomes that can be framed by this differentiation.

The perspectives, although distinct will be influenced by each other through the manner in which the environment is formulated. A differentiation can be established based on whether the environment (DP) is formed as an [evolution of] the description of a situation (R) or whether it is a projective imposition of the environment (DP) on a situation (R). A simplification of this distinction would relate to an inductive formation of an environment, in which the local environments, and consequently, their underlying worldview dominates perspectives, vs. a deductive formation in which the synoptic environment precedes, and consequently dominates the formation of local perspectives.

To resolve this duality, we require an algedonic enactment of R-DP or the maintenance of a participatory role. The determination of how action, if framed, will be defined by the underlying worldview. Irrespective of the worldview we can however state that any action is the execution of a select alternative of possible actions, and that any action is the enactment of a perspective, chosen or derived from a set of possible perspectives. This creates a sequential (not necessarily causal) construct between possibilities and actualities in which the future is merely the manifestation of one of the possible actualities, or vice versa, between actualities and possibilities in which the future is merely the manifestation of an actual possibility.

As an example: in the Schrödinger's Cat thought experiment (Schrödinger 1935), the two states in which the cat exists before the box is opened form the Realmoglichkeiten. This construct is generally arrived at through rational means, although empiricist approaches which are driven by listing of alternatives, such as brain-storming, threat analysis, scenario analysis or use cases will also generate a similar condition (although, unlike the rational construct these approaches are often unbounded or divergent). To determine an actual outcome, an action is necessary. The box must be opened, at which point we will see whether the cat is dead or alive establishing a Realwirklichkeit. The actuality may ultimately be influenced by the nature of the initial framing. The empiricist view will reflect an observed reality. This will be the more common of paradigms governing actualities. A rationalist view will state the reality as a function or manifestation of the rationalist framing and would take the form of a state or condition of the originally defined construct. This is typical in OR, theoretical work, chaos theory and other methods, where the actuality is merely a manifestation of the generalized construct from which it is derived; e.g. a subject to statement, or an outcome of a fractal.

Finding the cat dead or alive does imply that we could have known more before the box was opened, because until the very moment at which we looked in, the cat could have been either, or both. Given that we should be switching between underlying paradigms; from rationalist to empiricist, or vice versa, also implies that the Realmoglichkeit and Realwirklichkeit cannot simply be reconciled.

In general cases, where we have not one, but a quasi-continuous set of switches between the two spaces; the possible and the actual, a new baseline must be established after each collapse of the possibility function into an actuality; and consequently a new set of possibilities considered. This rapidly grows the set of extended possibilities (i.e. possibilities that are built over a series of iterations). The growth is exponential and similar to the logistic curve found in Chaos Theory. Metaphorically, it can also be represented by Heisenberg's Wave. The more dynamic and uncertain that a situation is, the more rapidly the wave diverges. In the construct that we are using this is analogous to the separation between DP and R. In such situations, the time interval between iterations must be reduced and new baselines established to manage the divergence in the dissipation of possibilities.

2.3.11 The Environment – Spatial Integrity

It would seem that under these conditions everyone functions in their own situation with little to share (even if willing) with others. Luckily, even though each individual is unique we share sufficient commonality for shared awareness to form. There are two mechanisms by which this is postulated to happen.

1. For cases where there is a high degree of uniformity a synthetic shared understanding is possible where through abstraction and aggregation a common perspective can be built.

2. For cases where there are significant differences a synoptic shared understanding is possible. The synoptic perspective is formed through a rational formulation of universal(s) within which component perspectives can exist, recognizing that other perspectives also form part of this synoptic construct, even though there is no specific overlap between individual components.

The formation of a shared perspective or understanding is fundamentally emergent. The complexity of the emergent process may however vary greatly.

In the simplest cases, the emergent shared perspective will be the same, or not significantly different, as the perspective of any of the components resembling what Bar Yam (1993) would refer to as local emergence. The degree of homogeneity between the salient features of the component perspectives would have to be significant with few or no relevant difference.

In cases where the component perspectives can still be deemed to be homogeneous, albeit with difference where the condition on the difference is that they be regularly distributed, a synthetic perspective will be possible. This synthetic perspective is in essence a form of sense-making reducing the existing variances in a manner that is congruent with the distribution. Depending on the stochastic nature of the variability, a local emergence may still be assumed.

It is when significant variance or heterogeneity between component perspectives is introduced that a form of global emergence (Bar-Yam 1993) will underlie the formation of a shared perspective or understanding. In these cases, the nature of the shared perspective cannot be known from a single or sample part of the components. To fully understanding the extent of the shared understanding, all of the components may have to be considered concurrently. Alternatively, the synoptically based shared perspective would allow for a preconceived shared perspective to exist that would govern the component understanding.

Understanding the manner in which shared (global) environments form, whether they be synthetic in nature or syncretic is an area of continued study. The emergence of shared perspectives is presently hypothesized to take place in a manner best described by percolation theory. According to percolation theory, a critical point is reached in which a phase transition occurs that would be representative of the formation of a shared transition. Present work is studying the effect of predisposition, proximity and intent and their effect on the critical point at which a phase transition occurs. The critical point is referred to as the K-Threshold; the point at which emergence becomes likely. This too is a meta-theoretical study, focusing on attributes of the situation, and the potential for emergence, without necessarily being able to say specifically what will emerge.

2.4 Research Methodological Considerations in Evolving ST

There are several thrusts defining the research in Situations Theory (Fig. 2.7, Table 2.2). These can broadly be categorized as (1) Ontological studies focusing on the nature of situations, (2) Methodological studies that draw from the understanding that ST

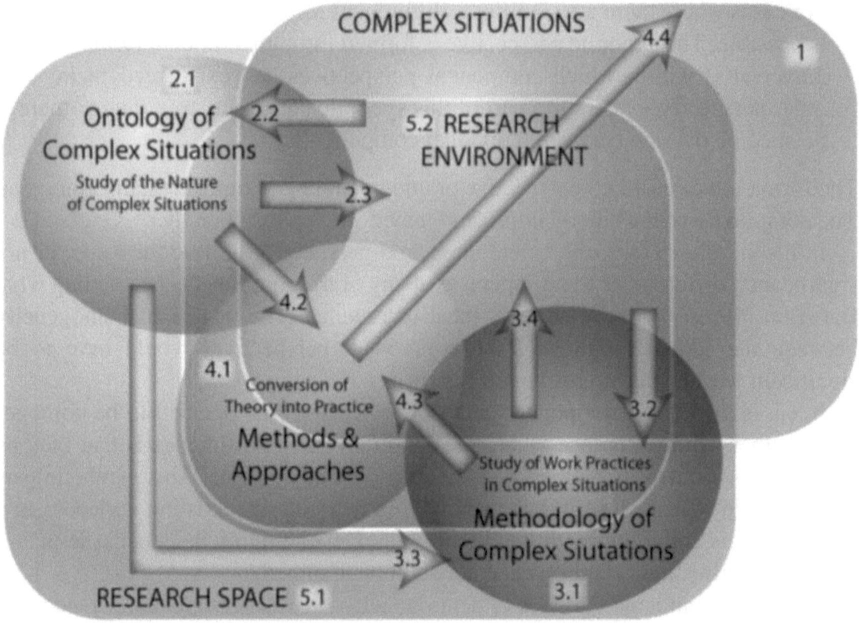

Fig. 2.7 The research methodology elements

Table 2.2 Description of the research methodology elements

1. Relevant wicked problems and other situations of interest
2. Complex situations (Ontology)
   1. Study of the nature of complex situations: this element proposes to study complex situations identifying their dominant characteristics and attributes, their behaviors and other ontological aspects
   2. The study of complex situations is moderated by the context encountered in the situation. The intent is to identify specific complex constructs that are faced in the situation [type] being studied
   3. *Influence on the practitioner: understanding the nature of the domain in which the individual must participate*
3. Work in complex situations
   1. Study of the work methods and best practices that are applied in complex situations. The general focus of this study encompasses any complex environment and includes documented findings of best practices
   2. Focus is on studying practices presently undertaken which may be compared to the general definition of best practices
   3. Best practices are compared and related to the specific nature of complex situations to better understand which methods, approaches and practices work under what conditions
   4. *Influence on the Individual: an improved understanding of the effectiveness of present practices, comparison of present practices to general best practices, and a decision making capability of identifying what practices to use under specific conditions*

(continued)

Table 2.2 (continued)

4. Methods and approaches
1. Fundamentally it is expected that new methods and approaches will be required to satisfy situations for which existing methods and approaches fail to provide effective solutions. The new methods and approaches will be informed by:
2. The generalized constructs of complex situations, and
3. Existing knowledge of best practices
4. *Influence on the Individual: through new practices extensive leveraging of the knowledge of complex situations and other capabilities is expected*
5. Research methodology
1. Research methods. Research of Situations and Situations Theory is multidisciplinary encompassing a variety of approaches including: Empirical techniques, experimentation, survey and correlational studies, grounded theory and inductive research, case studies and rational techniques (OR, M&S)
2. A Research Environment that captures empirical and experimentally generated data may support the research. The environment is intended act as both a test base for knowledge acquisition and as a learning space for knowledge dissemination. The environment provides and efficient interface in which learning can be undertaken to actively transfer knowledge to DHS. The environment will be capable of Live, Virtual and Constructed (L-V-C) training modes

provides and attributes and requirements that a situation might impose, and (3) development and evaluation of methods, tools and approaches as part of a pragmatic effort. These are meta-theoretical studies that elaborate on attributes of complex situations that are suggested to have an influence on practice and understanding.

From Sousa-Poza and Kovacic 2008

2.5 The Research Environment

The architecture for the environment is extendable allowing for a near seamless transition from

- A constructive environment in which a purely simulated laboratory configuration is used, to
- A virtual in which the simulated 'reality' is augmented by human players (HITL), to
- A live environment in which an empirically, rich human exercise is supported by both simulated and other technical components, to
- Integration and testing in a live operation.

The laboratory is designed on traditional experimental concepts expanded with fundamental theory (Fig. 2.8). The expansion of the traditional experimental designs with the RDP model (Sousa-Poza 2005) provides: (1) an advanced protocol for guidance on the design of the testing and decision environment to optimize the testing process, and (2) a metrification of confidence of the test or experiment that is run based on the conditions of the problem that is to be addressed.

The decision and testing environment is highly reconfigurable, and largely extendable, allowing for near seamless transition from one configuration to another,

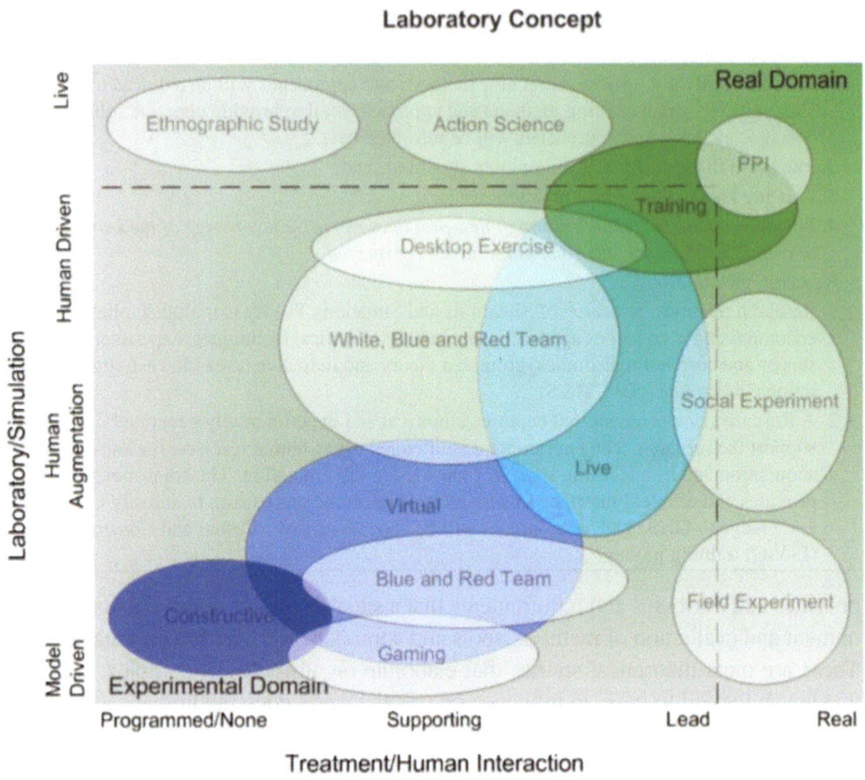

Fig. 2.8 The research "Laboratory"

as well as offering a high degree of compilation of runs executed in different configurations. The underlying support of the Research Platform provides a high degree of persistence to any sequence of tests or experiments, as well as an evaluative capability to assess the actual test construct.

Within the simulated environment, treatment may be applied following traditional experimental design approaches. In this layout a high degree of control is provided internal to the experimental environment, from which, in general, highly concrete, causal statements can be made of relationships within the laboratory. The trade-off, particularly for complex social environments, comes at the cost of assumptions and approximations that must be made in the development of the simulation. Conversely, a 'live' layout results in lower abstraction in the establishment of the laboratory with reality external to the laboratory. This, however, generally results in lower control within the environment reducing the concreteness of statements that can be made.

The ability to transition between layouts allows for a higher degree of confidence to be established in conclusions. The layout is ultimately expandable beyond the bound of the laboratory into live or 'real' tests in the field; with continued support of the research platform to ensure continued learning and adaptation.

2.6 Research Agenda

2.6.1 Micro Situations

Micro-Situations Theory focuses on the development of elementary understanding; understanding that is generated from perspectives.

Perspectives are cognitive constructs and ultimately the purview of an individual. Every individual's perspective of a situation is fundamentally unique; which is not to say that there are no aspects of perspectives that are not shared by two or more perspectives. Micro-Situations constructs focus on the development and evolution of perspectives as well as the different constructs that explain the manner that shared constructs or perspectives are both formed and communicated. Both mechanisms, the formation and the sharing of perspectives, have multiple models that explain how these occur and are to be understood. Not all of the constructs provide for the same conclusions.

Different models of the formation of perspectives can be found in literature, including the lens model, explanations involving worldviews, and others. Fundamentally, most of them rely on an element of 'what is real', a conversion of that reality by an observer, and the ensuing perspectives. Focus lies heavily on the conversion process and the nature of the perspective. The conversion constructs vary in nature depending on the school of use. In positivist thinking, for example, the conversion constitutes a 'bias' and has pejorative connotations; in constructivist terms on the other hand, the conversion is fundamental, inescapable, and necessary for the interpretative process that underlies the formation of perspectives.

Sharing has less to support it, although concepts such as 'common operating picture', shared or common awareness, collective learning and organizational knowledge are widespread in use. The idea of sharing is generally an imposed and often relatively positivistic concept, assuming that that which we see is that which is. What a shared perspective is (whether it is a global construct or elements of perspective that are common across individuals) is never clearly delineated. Furthermore, there is little investigation on synthetic perspectives (perspectives that provide aggregate general insight of all perspectives) or synoptic perspectives (perspectives that incorporate all perspectives but provide little insight on any specific perspective), even though their use is not uncommon or undocumented.

Micro-Situations Theory provides a set of concepts, models, processes, and 'ideas' on these important constructs.

Micro-Situations Theory rests on a particularistic assumption that ultimately stipulates that the situation is: (1) an enactment of the collective existence of perspectives, and (2) that the efficacy of a situation is contingent on the effectiveness in the formation of perspectives, the breadth in the types of perspectives, and the degree to which these are able to 'co-exist'.

2.6.2 *Macro Situations*

Our understanding of macro-situations theory is less evolved than the micro-situations constructs. In part this is due to the manner in which the research began, which was focused initially on elementary constructs surrounding knowledge and understanding. A lot of effort has consequently gone into this line of work. Micro constructs are also the basis on which self-reflection takes place, an exercise that we have all gone through extensively in trying to both separate ourselves from existing dogma, and trying to break from personal orthodox views. It is, however, also due to the limited opportunity to study collective constructs, although some [limited] opportunity has been available through reflection of our own progress as a group.

Macro-Situations Theory deals with the behavior of whole, studying the collectives as universal constructs. Fundamentally Macro-Situations Theory still rests on the tenet that it is understanding that leads to action; and that the qualitative nature of understanding impacts the quality of action. It also rests on the concept that the Representation (as a Domain-Perspective or DP construct) is in essence a separation from Reality (R). The similarities with Micro-Situations Theory, however, end at this point.

Macro-Situations Theory is focused at identifying practical global processes and activities that maximize the overall understanding within a situation. In extremely simple situations where the component perspectives are extremely homogeneous, there is one (or very few) component perspectives; the overall understanding is fundamentally equal to the understanding of a single component. In the general condition, however, the overall understanding cannot be stated as a function of the individual components (although it is effectively composed of these elements). To complicate matters, there are multiple theories that can be applied to derive a statement of understanding.

Macro-situations constructs are fundamentally universalistic in nature, where the concept of the environment precedes the formulation of individual perspectives. Efficacy within this construct indicates that effectiveness is governed by the ability of the environment to create a shared or common domain and provide for means to allow for individual perspective to co-exist within this structure.

2.7 Imperative of Situations Theory: The Call to Heretics

Situations Theory is not about solving problems but rather about understanding the paradigms that are applied to study and solve problems. The intent is to study what works and what does not, and to understand why this is the case. Through this understanding, it has become possible to gain a unique insight into [complex] situations such as those that Rittel and Weber (1973) describe.

That is not to say that "solutions" have been found, but rather ways to think about the complexities have been elaborated. The utility of understanding the nature of a situation better lies with the improvements that have been, and might be made, in dealing with the a specific situation. At very least, it should reduce senseless activity in which it is clear that there is no feasible manner to achieve a desired outcome.

Insanity: doing the same thing over and over again and expecting different results

Albert Einstein

It is interesting, however, how strong the pushback is when a challenge is made to the worldviews and paradigms being applied. For many it would appear preferable to continue down a fruitless inquiry than to question the reason and methods being applied. Ultimately, this may be due to the manner in which a paradigm is engrained, not only in how we think, but also possibly in defining who we are.

"There is nothing louder than the sound of a breaking paradigm"

Charles Keating

Given the complexity of a situation is contingent on the observer and the manner that situation was generated, there is a very strong likelihood that the paradigm underlying the generative process is partially responsible for the complexity and difficulties encountered in understanding the problem. Dealing with the systemic intricacies that might be present in an engineering or organizational problem is often necessarily accomplished through the implementation of highly evolved technologies, processes, procedures and structures. There comes a point however, as Ray Ozzie pointed out, that this "complexification" of the solution becomes its own worst enemy. At this point it becomes imperative to change the dialogue, the very manner in which the problem is being addressed.

Any intelligent fool can make things bigger and more complex... It takes a touch of genius - and a lot of courage to move in the opposite direction

Albert Einstein

This is not always easy, particularly if individuals have a strong vested interest in the work being done.

Over the millennia there have been many of Gladwell's (2000) 'tipping points' that have been encountered, at which a change in thinking became necessary. Art Kleiner (1996) refers to these instances where change was undertaken as the age of heretics, making reference to Pelagius who challenged 5th century church doctrine with his ideas of free will and individual morality. Pelagius was confronted by Augustine (who was canonized to sainthood, should there be any doubt who held sway in this debate). There seems to be a degree of truth that change is seldom embraced!

"History warns us, [...], that it is the customary fate of new truths to begin as heresies and to end as superstitions"

Thomas Henry Huxley

Huxley also warns that any truth, once accepted, will become the new dominant discourse that will no longer be questioned. This cycle is not new.

We often fail to see that which is apparent, only to then fail to see it once it become commonplace.

This is not a matter of creating new technologies, artifacts, or solutions. It is a matter of questioning what is perceived; the situations that are encountered. It is quite likely that what is being sought is already available; the next great breakthrough has possibly already been made yet remains to be "seen". After all, Heron of Alexandria invented the steam turbine in 100 AD, 1,500 years before it was rediscovered in the industrial age.

Change in management styles has also required its revolutions. Frederick Winslow Taylor (1947), father of scientific management, was called to testify on his methods before a Special House Committee. Given the gains that organizations and employees were making it seems inconceivable that this should have taken place. In retrospect, we may have succumbed to complacency in questioning. Irrespective of the gains that scientific management has provided, some of the criticisms that Taylor faced on the problems of overspecialization, and "mechanizing" employees may now be surfacing as detrimental in a knowledge-based economy. This does not make scientific management wrong, but rather it must be recognized that it may not be appropriate for all problems.

Of course, only history will tell whether we are at a 'tipping point' of management theory, and what the outcome will be.

References

Bar-Yam Y (1993) Dynamics of complex systems. Westview Press, Boulder

Bertanlanffy V (1976) Ludwig, general system theory: foundations, development, applications. George Braziller, New York

Bohr N (1949) Discussions with Einstein on epistemological problems in atomic physics. In: Albert Einstein: Philosopher-Scientist. Cambridge University, Cambridge

Brewer Van A (2010) Decision-making construct for complex situations. ODU Dissertation, May 2010

Brewer V, Gheorghe A (2011) Forensic complexity. Int J Syst Syst Eng 2(4):347–355

Brewer V, Sousa-Poza A (2009) Generalized research canons for JTB(+) knowledge, international symposium on peer reviewing (ISPR 2009). In: Proceedings of the 13th world multi-conference on systemics, Cybernetics and informatics (WMSCI 2009), Orlando, July 10–13, 2009

Checkland P (2006) Systems thinking, systems practice, 2nd edn. Wiley, New York

Conklin J (2006) Wicked problems and social complexity. In: Dialogue mapping: building shared understanding of wicked problems. Wiley, Chichester

Dauer F (1974) In defense of the coherence theory of truth. J Philos 71(21):791–811

Gladwell M (2000) The tipping point, how little things can make a big difference. Little Brown Publishing, Boston

Hartman N (1949) New ways of Ontology. (trans: Reinhard CK). -Henry Regnery Company, Chicago, 1953, pp 13–14

Keating C, Rogers R, Unal R, Dryer D, Sousa-Poza A, Safford R, Peterson W, Rabadi G (2003) System of systems engineering. Eng Manag J, Rolla, MO, 15(3):36–45

Keating C, Sousa-Poza (2005) A system of systems engineering analysis of integrated port security for the Virginia port authority. NCSOSE-TR-002-05. Agency: Virginia Port

Authority, Number of Figures: 15, Number of Tables: 5, Number of References: 29, Pages: 125, June 2005
Kleiner A (1996) The age of heretics. Doubleday, New York
Kovacic S, Sousa-Poza A, Keating C (2006) Complex situations: an alternate approach for viewing a system of system. IEEE/SMC Proceedings
Kovacic S, Sousa-Poza A, Keating C (2007a) The national centers for system of systems engineering: a case study on shifting the paradigm for system of systems. Syst Res Forum 2(01): 52–59. World Scientific Publishing Company, Singapore
Kovacic S, Sousa-Poza A, Keating C (2007b) Type III: the theory of the observer. SoSE'07. In: IEEE international conference on system of systems engineering. IEEE, New York
Kovacic S, Sousa-Poza A, Charles Keating (2008) Complex situations: an alternative approach for viewing as system of systems. In: Proceedings of the third international conference on system of systems engineering, Monterey, CA, June 2–4 2008
Millikan RG (1986) The price of correspondence truth. Noûs 20(4):453–468
Nonaka I, Konno N, Toyama R (2001) Emergence of "Ba": a conceptual framework for the continuous and self- transcending process of knowledge creation. In: Nonaka I, Nishiguchi T (eds) Knowledge emergence. Oxford University Press, New York, pp 13–29
Oizerman TI (1981) Kant's doctrine of the "Things in Themselves" and Noumena. Philos Phenomenol Res 41(3):333–350
Padilla J (2007) Non-monotonic synopsis: an approach for assessing complex situations. In: MECS forum conference proceedings, Norfolk, pp 182–191
Padilla JA (2010) A theory of understanding within problem situations. ODU Dissertation
Padilla J, Sousa-Poza A, Tejada A, Kovacic S (2007) Towards a Diagnostic Framework for Understanding Complex Situations. In: NECSI conference, Quincy, MA, 28 Oct–2Nov 2007
Padilla JJ, Diallo SY, Sousa-Poza AA (2011) Establishing a theoretical baseline: using agent-based modeling to create knowledge. In: Tolk A, Lakhmi J (eds) Intelligence-based systems engineering, vol 10. Springer, Heidelberg, pp 259–284
Rittel H, Webber M (1973) Dilemmas in a general theory of planning, policy sciences, vol 4. Elsevier Scientific Publishing Company, Amsterdam, pp 155–169
Schrödinger E (1935) Die gegenwärtige Situation in der Quantenmechanik. Naturwissenschaften 23(49):823–828
Skyttner L (1999) Introduction to general systems theory. Plenum, New York
Sousa-Poza A (2007) Plenary presentation. MECS Conference, Norfolk, VA
Sousa-Poza A, Correa Y (2005) Pragmatic idealism as the basis for understanding complex domains: the trinity and SOSE. In: Proceedings of the IEEE SMC 2005 conference, Hawaii, 10–12 Oct 2005
Sousa-Poza A, Keating C (2004) An investigation into the use of recursion in system of systems engineering. NCSOSE-TR-002-04. Agency: concurrent technologies corporation, Number of Figures: 2, Number of Tables: 1, Number of References: 21, Pages: 32, December 2004
Sousa-Poza A, Kovacic S (2008) A research agenda for complex situations. Eng Manag J 20(4): 32–39
Sousa-Poza A, Keating C, Kovacic SF (2004) System of systems analysis of the maritime security system for the port of Hampton roads. NCSOSE-TR-001-04. Agency: US Coast Guard and Concurrent Technologies Corporation, Number of Figures: 13, Number of Tables: 21, Number of References: 29, Pages: 123, April 2004
Sousa-Poza A, Keating C, Mun JH (2004) System of systems methodology. NCSOSE-TR-001-05. Agency: Department of Homeland Security, Number of Figures: 8, Number of Tables: 3, Number of References: 16, Pages: 53, April 2004
Sousa-Poza A, Kovacic S, Keating C (2008a) System of systems engineering: an emerging multidiscipline. Int J Syst Syst Eng 1(1/2):171–188
Sousa-Poza A, Padilla J, Bozkurt I (2008) Implications of a rationalist inductive approach in system of systems engineering research. In: Proceedings of the third international conference on system of systems engineering, Monterey, CA, 2–4th June 2008

Stacey RD, Griffin D, Shaw P (2000) Complexity and management: fad or radical challenge to systems thinking?. Routledge/Taylor & Francis Group, London

Taylor FW (1947) Scientific management (comprising shop management, the principles of scientific management, and testimony before the special house committee). Harper and Row, New York

Trans-Atlantic Research and Education Agenda in Systems of Systems (T-AREA-SoS) (2012) "TAREA-PU-WP2-R-LU-9, SOA Report", © Loughborough University

Unites States Air Force Science Advisory Board (2005) Systems of systems engineering for Air Force capability development", SAB-TR-05-04, July 2005

Chapter 3
PRISM – A Philosophical Foundation for Complex Situations

Van E. Brewer

Abstract The uncertainty inherently associated with complexity challenges decision-making processes, indicating a need for a construct for decision making in complex situations. A review of the literature on systems, complexity, and paradigms indicates that such a construct must be internally consistent with well-defined philosophical foundations and further that systems and complexity (as used in complex situations) are not necessarily internally consistent with traditional philosophical foundations. Therefore, a decision making construct for complex situations requires research into different foundations. This research addresses these gaps, deriving axiological and methodological components based on a set of principles consistent with the ontology and epistemology of Sousa-Poza and Correa-Martinez (Pragmatic idealism as the basis for understanding complex domains: the trinity and SOSE. In: 2005 IEEE international conference on systems, man and cybernetics vol 3, pp 2744–2750, 2005). The combination of these four philosophical components – a pragmatic ontology, rational epistemology, idealist axiology and situational methodology – is asserted to establish a foundational perspective for complexity and systems.

Keywords Complexity • Complex Situations • Decision Making • Knowledge Understanding • Justified True Belief • Pragmatism

Complexity presents itself in various forms such as chaos, wicked problems, nonlinear dynamics, and soft systems, and arises from diverse sources such as mathematics, systems theory, dynamics, and biology. However it is manifested, complexity reveals a fundamental characteristic of uncertainty, which in turn brings to mind the foundational aspects of certainty – knowledge, reality, and the methodologies used

V.E. Brewer (✉)
Department of Engineering Management and Systems Engineering
Old Dominion University, Norfolk, VA, USA
e-mail: van.e.brewer@gmail.com

S.F. Kovacic and A. Sousa-Poza (eds.), *Managing and Engineering in Complex Situations*, Topics in Safety, Risk, Reliability and Quality 21, DOI 10.1007/978-94-007-5515-4_3,

to relate the two. Traditional epistemological and ontological constructs inevitably lead – under conditions of uncertainty – to conditions which are inconsistent with their associated philosophical foundations. (Sousa-Poza and Correa-Martinez 2005; Keating et al. 2005; Brewer 2010) and others have initiated a dialog exploring constructs for complex situations, framing the discussion around the elements of epistemology, ontology and methodology which carries potential to enrich research and practice.

The Pragmatic-Rationalist-Idealist-Situational Model,[1] or PRISM, provides a philosophical foundation based on the ontology and epistemology of Sousa-Poza's Pragmatic Idealism (2005), deriving axiological and methodological components based on a set of principles consistent with that work (Brewer 2010). The combination of these four philosophical components provides an internally consistent foundation for complexity; including, when paired with the idea of generalized canons (Brewer and Sousa-Poza 2009), a consistent set of research canons based on a specific definition of knowledge. The construction of this model is helpful to understand some basic principles as fallibility and comprehensibility, which permits the purposeful construction of methodology (in the sense of a means to guide in the selection of methods). Ultimately, PRISM is a foundation that permits a consistent consideration of complexity – a fundamental element in building methods to operate in such conditions.

3.1 Complexity and Understanding

The PRISM foundation uses the definition of complexity from Sousa-Poza and Correa-Martinez (2005): complexity is "proportional to the probability of having/making an erroneous knowledge claim" (section 3, para. 2). This probability of erroneous knowledge claims directly impacts understanding; indeed, complexity as "a construct associated with the fallibility of understanding" (section 3, para. 2). It remains, then, to examine understanding and assess the implications of complexity (as so defined). A standard definition of understanding is "to comprehend; to apprehend the meaning or import of; to grasp the idea of" (OED 2008). Dewey (1933) speaks to the same concept but is more explicit in the impact of understanding: "To grasp the meaning of a thing, an event, or a situation is to see it in its relations to other things: to see how it operates or functions, what consequences follow from it, what causes it, what uses it can be put to. In contrast, what we have called the brute thing, the thing without meaning to us, is something whose relations are not grasped. … The relation of means-consequence is the center and heart of all understanding" (pp. 137, 146). From this perspective, higher complexity implies a lower probability of establishing this means-consequence relation – causality is lost, and the ability to understand is limited in kind. Also, understanding rests with the observer; hence complexity in a

[1] This model was initially developed as a "Complex Situations Paradigm" in (Brewer 2010). Though the label of a paradigm was strictly associated with the definition of a paradigm as composed of four philosophical components, further exploration of the results indicated that a Kuhnian paradigm shift can occur within the "paradigm". As a paradigm shift within a paradigm is confusing at best, the moniker has adapted, accordingly, to simply one of a "model".

situation is linked to the observer and the observer's understanding of the situation. An understanding of complexity permits the observer to place a low understanding of the situation in context, establishing a different form of means-consequence relation for a complex situation where the lack of causality is understood to be a characteristic of the situation. From this perspective, understanding a complex situation becomes not a paradoxical statement but a statement that incorporates the limits of understanding. Complexity in a situation therefore directly influences understanding of that situation, requiring the observer to understand complexity in order to understand the situation in a tractable context. Understanding is related both to knowledge and the ability to place that knowledge in context; hence an implication of complexity is a limit to understanding of the situation. The concept of understanding a complex situation must be refined and ultimately supplanted with the concept of understanding complexity. The observer is faced with a situation where the statement *I don't know* (with respect to the complex entity) has been replaced with *I can't know*, certainly a critical concept to understand.

Treating understanding (and by implication knowledge) as the underlying principle behind the diverse lineage of complexity, it is immediately noted that a major shift in perspective has occurred – the focus has moved from the monolithic concept of a complex thing (or collection of things) to include the observer. Drawing on the complex situation construct established by Sousa-Poza and Correa-Martinez (2005), the observer is inextricably bound to the complex situation, and so establishes a reference point for assessing perspectives on complexity. The impacts on the observer as a reference point can then be assessed with respect to this underlying principle. This explicit introduction of the observer introduces the requirement for a strong analysis of philosophical foundations, including such fundamental issues as epistemology and ontology. These discussions are often assumed in engineering subject matter areas where positivist empirical approaches are dominant in the discourse – with some notable exceptions (e.g. Heylighen et al. 2007; Georgiou 2000; Sinn 1998; Sousa-Poza and Correa-Martinez 2005). The latter proposes pragmatic idealism, exploiting ideas of Rescher (1994, 2000) to propose an ontological foundation for exploring complexity. Due to this critical emphasis on philosophical foundations for discussions involving the observer, this model endeavors to make these philosophical assumptions explicit. *Complexity contradicts the assertion that more knowledge results in more understanding*; *it challenges the notion that reality is fundamentally comprehensible.* This introduces the concept of fallabalism, which can be taken to be a justification for inaction (as no action can be rigorously supported); but as this discussion presumes a value premise driving action, the perspective of pragmatic idealism (Sousa-Poza and Correa-Martinez 2005) is appropriate.

3.2 Philosophical Foundations

The essential elements of a philosophical foundation can be interpreted as epistemology (study of knowledge), ontology (study of being), and methodology (study of relating being, or reality, to knowledge). However Lehmann (2004) provides an interpretation that includes "the praxis for applying the theoretical constructs in epistemology into behavioral codes of conduct" (pg. 19). While the branches of philosophy are many and varied, and can be debated ad infinitum, this distinction is of some import as it shifts the emphasis on *how* to the ethics that form the basis of how. From this perspective, the notion of methodology in isolation is sterile absent a reason or goal for the methodology, so this discussion extends to include an axiological component.

3.2.1 Epistemology

In the simplest epistemic perspective, the opposition of rationalism to empiricism is in the source of knowledge: it is either derived from reason or from experience – and in the extreme sense, it is derived solely from one source or the other. These positions are echoed in the philosophical perspectives of positivism which in the purest sense holds all knowledge to be gained through experiential means, and constructivism which in the purest sense holds all knowledge to be constructed within the intellect. A critical discussion of constructivism is found in Olssen (1996), which places it in the context of realism.

This debate of the rational vs. the experiential is traceable to positions set forth by Plato and Aristotle. Plato (1892) held that ideas (*eidos*, or Forms) are real and universal and matter is a specific instantiation of a form. Therefore *treeness* is real and the *tree* is an instance of treeness – agreeing that a thing is a tree constitutes implicit agreement on the treeness of the thing. This hints at an underlying assumption of the duality of the material and immaterial – or body and soul – within Platonic philosophy; the mind is seen as distinct from reality and therefore perceptions within the mind have a fundamental reality. In a sense, knowing is unique and real, and the known is an instantiation of knowing – and therein lies the root of rationalism and constructivism. Concisely phrased, reality is a subset of knowledge. In this perspective, the presence of the duality is central to the construct. Conversely, Aristotle (1994) submits that a particular substance is a combination of both matter and form; in this view, universals exist only as they are instantiated. Therefore, the *tree* is real, and *treeness* is a quality of the tree. The essence of the argument is the existence of a unitary objective reality; something that knows is part of this reality, hence knowing is subsumed by reality. In concise phrasing, knowledge is a subset of reality. The diametrically opposed conclusions of the two philosophical perspectives reflect back to the notion of internal consistency with philosophical foundations. A perspective that rests upon a duality (whether material/immaterial or mind/body) permits subsuming one to the other; a perspective that rests on a unity necessarily subsumes all to that unity.

The comparison across the extremes of rationalist and empiricist epistemology represents only a fraction of the overall epistemic debate. Rosenberg (2005) provides an in-depth summary of the history of this dialog with respect to the philosophy of science; of particular interest is the assertion that empiricism has a preeminent role as "the 'official' epistemology of science" (pg. 23). This dialog evolved into post-positivism or postmodernism, emphasizing (in various forms) the role of the observer in not only the collection of data and generation of knowledge, but the development of the questions that frame inquiry. The introduction of this degree of epistemic relativism poses a serious challenge to a rigorous definition of knowledge generation – as Popper (1934) stated, "We do not know, we can only guess" (pg. 278). This indicates the need for significant caution and clarity when departing from the (asserted) assured reality of empiricist epistemology. As pragmatic idealism (Sousa-Poza and Correa-Martinez 2005) introduces both the observer and the solution form (the form of interaction) in the complex situation, this degree of rigor in the generation of knowledge claims is essential for the proposed research. van Uden et al. (2001) and Cilliers (1998) also explore the relationship between postmodernism and complexity; in these cases they tend to reduce the uncertainty to an application-focused level (e.g. more agile, more flexible, more responsive) that displays a transformative teleology within a formative environment (Stacey 2000).

Finally, Azzouni (2004) provides a set of epistemic conditions for observations that reflect a "tracking" requirement. This reflects the interdependency between epistemology and ontology: "The epistemic processes, which establish truths that we're committed to, must be sensitive to the objects *about which* we're establishing those truths" (pg. 372), which reinforces the rational requirement for consistency across the elements of the philosophical foundation.

3.2.2 Ontology

Turning to the ontologic component, general approaches to the nature of existence tend to focus on the differentiation between an objective reality (realism) and a subjective (or constructed) reality (non-realism, under various forms such as postmodernism). Miller (2005), Feyerabend (1981) and Vandenberghe (1999) outline the debate in some detail, the latter being from the perspective of sociological theory. In this instance, a more precise focus as defined by Rescher (2000) is used. The pragmatic assertion is that of a mind-independent reality in order:

1. To preserve the distinction between true and false with respect to factual matters and to operate the idea of truth as agreement with reality
2. To preserve the distinction between appearance and reality, between our picture of reality and reality itself
3. To serve as a basis for intersubjective communication
4. To furnish the basis for a shared project of communal inquiry
5. To provide for a fallabilistic view of human knowledge
6. To sustain the causal mode of learning and inquiry and to serve as the basis for objectivity of experience (pg. 100).

Note that this preserves aspects of realism and non-realism, in a manner similar to postmodernism. These distinctions provide a sense of direction for subsequent development. Having established the distinction for the observer, pragmatic idealism rejects fallabalism not in the sense that error can be present, but in the sense that action must still be taken (e.g. a decision must be made). Further, the notions of causality and learning are directly related to knowledge generation, and indicate that this foundation can be related to understanding. Finally, there is an emphasis on intersubjective communication and communal inquiry that indicates the need to address multiple actors involved in knowledge generation.

3.2.3 Axiology

Axiology (from the Greek *to speak of values*) is a branch of philosophy addressing the study of value (OED 2008). Value is "the relative status of a thing, or the estimate in which it is held, according to its real or supposed worth, usefulness, or importance" (OED 2008). Value theory therefore has a significant breadth of subject matter: "acts or processes of mind involved in our responding, appraising, and judging and in the creation of values … concepts we make use of in these acts or processes of evaluation … the presentation of completely general and ultimate apprasive concepts such as good" (Aschenbrenner 1971, pg. 4). The breadth of inquiry is further illustrated by Jessup (1949): "Among writers on value there seems to be fair agreement that these things include: (1) value-feelings (pleasures and pains) and affective dispositions; (2) volitional activity (based on felt tensions), setting up ends or purposes, the drive toward which is not simply additive of the moments or points of felt-pleasures along the way; (3) value-judgment (appraisal or evaluation), which may reject, censure, or approve of both feeling and will; (4) the value objective, which may be any thing, quality or situation, actual or ideal, and of any degree of intricacy" (pg. 127). Keeney (1992) provides a treatment of the role of values in decision-making, contrasting value-focused thinking to alternative focused thinking – in particular focusing on the ability to create decision opportunities. Notwithstanding the breadth of inquiry for value theory as illustrated in the above and in Hart (1971), elements of interest for this discussion include those related to being, knowledge and praxis (Lehmann 2004).

The notion of value necessarily requires a subject and a predicate, or a judgment and a subject of the judgment – both valuing and valued entities are presumed. These are essential ontological assertions towards the existence of each; however, no further assertions are required at this level of definition. Whether a rationalist, empiricist, realist, pragmaticist or other ontological foundation is asserted, all can result in valued and valuing entities. Rational foundations may assert entities derived through reason; empiricist foundations may assert an experiential basis for the same. In a broad sense, then, axiology is dependent upon an ontological foundation – whether it is derivative of this same foundation remains to be examined.

The epistemological foundations for axiology are clear in the sense that knowledge of the thing valued is available (though again the specific philosophy may vary). However, from the descriptions quoted above there are additional aspects to be examined – the above quotations make clear reference to value as both individual and situational. That there is even a debate on values indicates there is variation across individuals – the concept of subjective expected utility is a mechanization of this variation. Either values are universal and constant and the discovery of these values differs across individuals, or values are individual and situational and consequently differ across individuals. Therefore, as with the earlier ontological argument, axiology will be dependent upon and derivative of an epistemological foundation. The extent of this dependency is significant as the discussion of value requires a discussion on the nature of self-awareness. That there is a need to choose indicates a reason to choose – i.e. there is some purposive end for the individual making the decision. Though the dialog on values (and decisions) deals with far more pragmatic ends, *at the heart of the dialog there is an assumption of an ultimate purposive end that creates the need for a self-aware individual to make a choice.* Further, this requires the ability to create some representation of alternatives, and discern some benefit towards the ultimate purposive end. These ideas are implicit in value theory and must be accounted for in the supporting ontological and epistemological foundations.

In the extreme, this becomes a debate as to whether axiological foundations exist independent of self-awareness. In the total absence of self-awareness, there is by the above arguments no valuing entity, and the debate is reduced to the notions of value independent of valuation – an inconsistent concept unless the ontological foundation asserts value as an inherent characteristic of existence, one that is discovered in the presence of self-awareness rather than assigned. Hence there are two cases to consider: either value is assigned by the valuing entity, in which case there is an ontological assertion of existence and self-awareness and a derivation of value through epistemological assertions; or it exists as an ontological assertion and is revealed through an epistemological assertion. The former presents a constructivist perspective, the latter a positivist perspective – but in both cases it can be asserted that axiology is not only dependent on ontology and epistemology, but derivative of them as well. As methodology has already been presented as requiring a reason (value) for application, the complete suite of a study of being, knowledge, values (derivative of the same), and praxis (methodology) then forms a coherent structure for a philosophical foundation.

Understanding implies a sense of causality – the "means/consequence relationship" (Dewey 1933, pg. 146) – that is at odds with unpredictability. Under uncertain circumstances, behaviors are stochastic whereas in unpredictable circumstances, behaviors are neither stochastic nor deterministic. The adoption of a time-independent perspective as opposed to a time-dependent perspective does not necessarily result in any change in the probability of an erroneous knowledge claim within the perspective. Unpredictable behaviors, for example, may be chaotic (deterministic but having such sensitivity to initial conditions that future outcomes cannot be determined to an arbitrary degree of accuracy) or complex (emerging from rich interactions of multiple entities). Further, complex behaviors are sensitive to history – every instance is unique and will have a unique associated future. In such cases it is not possible to quantify consequences – e.g. assign a probability to outcomes – as any such assignation has as little or as much meaning as any other. The selection of a given choice becomes irrelevant under such circumstances; simply put if it is not possible to know what outcome will occur in the instance or over time, it is not possible to choose between outcomes.

In such conditions, notions of assigning value to outcomes may seem moot as nothing can be done with the assigned value. However, it is appropriate to consider that considerations of value may return to a more fundamental discussion on the reason for making a choice – the notion that there is a purposive end. This must be phrased in a manner that incorporates those cases where consequence can be defined (e.g. there is a prior distribution) and where value can be assigned (e.g. there is a utility function). As consequence and value have been previously asserted to be based on some selected philosophical foundation, any consideration must be internally consistent with those foundations; hence the basic assertion is that an action or decision is justifiable if it is internally consistent with philosophical foundations. Here, justification is used in the sense of "To show or maintain the justice or reasonableness of (an action, claim, etc.); to adduce adequate grounds for; to defend as right or proper" (OED 2008). Rather than serving as an axiomatically specific term, it instead captures the conceptual subtleties of operations in unpredictable circumstances where the basis for choice is not clear – certainly not in the traditional sense, and potentially not clear in any sense. This is reflected by Aliseda (2004): "explanations are public objects of 'justification', [*sic*] that can be checked and tested by independent logical criteria" (pg. 341), and further by Rescher (2009a): "The prime directive of cognitive rationality is to maintain consistency" (pg. 3). The logical criteria are provided by the philosophical foundations – a rational deductive approach.

Finally, this must be stressed as a necessary axiological principle but it is by no means asserted to be sufficient – there will almost certainly be additional axiological considerations in every particular situation (for example, ethics, morality, etc.) that are not treated here. The justification principle is highlighted here as particularly relevant to this discussion, for actions both in research and in practice must be similarly justified. Therefore internal consistency, including a robust axiology, is a requirement for both research and practice.

3.2.4 Methodology

Kieseppä (2000) presents methodology as one of the principle consequences for epistemological and ontological choices. Mingers (2003) provides a summation of methodology as the general study of methods of intervention or research (as in a course in Research Methodology); the particular methods used in a specific project or study (as in "what was your research methodology?"); and a generic combination of methods that is commonly used as a whole (as in soft systems methodology, strategic options development and analysis, or survey methodology covering the design and analysis of questionnaires) (pg. 559). Methodology is distinct from methods; for example the methodology traditionally associated with logical positivism is the process of empirical scientific inquiry. Popper (1934) regarded methodological rules as conventions – the "rules of the game of empirical science" (pg. 53) (which indicates the association held with methodology at that time). Development of a methodology cannot focus on rules, but the way in which rules are to be generated. To speak of a *quantitative* or *qualitative* methodology is not necessarily consistent as these presuppose a specific method – not the guidance for generating specific instantiations of a method.

Further, Guba and Lincoln (1989) emphasize the growing tendency to shift "across paradigms" (p. 157) in methods. This raises serious issues with the notions of internal consistency; of course, it is not necessarily an issue when moving between methods that are consistent with a given methodology – for example, both quantitative and qualitative approaches can be consistent with empiricist epistemology. This tendency cannot be held to be true across diverse epistemic foundations; again caution and clarity is demanded. Axiological and methodological components of the foundation will naturally have a focused relationship; Lauden (1986) notes that "The methodology of inquiry has to be supplemented by the axiology of inquiry" (pg. 351). In practice, this indicates the need for co-development of these elements of the foundation.

A methodology may well be seen as an *a posteriori* meta-construct for a collection of methods developed for inquiry within an overall philosophical foundation, which provides little guidance for *a priori* development. However, it can be framed from the definition focusing on the "rules" for scientific inquiry, which leads to the assertion of the tautological relationship that a *methodology must provide guidance for the selection of methods for generating knowledge within a given philosophical foundation.* The reference to methods for generating knowledge immediately brings to mind canons for research. These are generally used as conventions for the design and execution of research, but the requirement is highly similar and thus suggests a close relationship between methodology and research canons. This also incorporates a subject that is not often explicitly treated in research methodologies – the definition of knowledge itself. From this perspective, development of a methodology can use the other elements of the philosophical foundation and research canons to facilitate selection or generation of methods for research.

As further guidance for the initial direction of the model, given that the methodology must be developed in concert with the other elements of the model, it may be possible to exploit known relationships between those other elements. An example of this can be found in (Rescher 1998) illustrating the tension between the "ontological and epistemological dimensions" (pg. 114):

[Ontological]	[Epistemological]
Unexplainable	Not (yet) explained
By chance	By some cause we do not know of
Spontaneous	Caused in a way we cannot identify
Random	Lawful in ways we cannot characterize
By whim	For reasons not apparent to us (pg. 114)

3.3 The PRISM Philosophical Foundation

It should be noted that, though this discussion provides the ontology, epistemology, axiology and methodology in that order, in development they are nearly. The initial development was framed around ontology, which culminated in a difficulty to resolve epistemology. Likewise, a subsequent effort to frame the effort around epistemology led to challenges in ontology and a difficulty in resolving a methodology. Ultimately, all perspectives are unique and can only be re-solved through a shift to another paradigm or worldview – and though the following discussion is necessarily linear, the end result is necessarily a holistic product as threads of ontology, epistemology, axiology and methodology are woven into an internally consistent model.

3.3.1 Ontology

The basic ontological principle is the existence of a "mind-independent reality" (Rescher 1994):

O_1: Reality is (that which exists). (Reality Principle)

This is consistent with Sousa-Poza and Correa-Martinez (2005), which asserts that "a reality exists as a construct, which is both separate and part of the observer, and is beyond the observer's full understanding" (Sect. 5.1, which brings an essential clarification to Rescher's statement as reality is framed as mind-independent and yet mind-inclusive. In this construction, reality (that which exists) exists and, in and of itself, is irreducible and cannot be subdivided or simplified – any division would establish another reality, the existence of which would extend reality. Here, irreducible is used in the sense of "… cannot be reduced to a simpler or more intelligible form; incapable of being resolved into elements, or of being brought under any recognized law or principle" (OED 2008). As such, it speaks solely to the subject of existence, and does not prohibit or require reducibility in other senses (e.g. it is possible to posit the existence of something "beyond reality," including such individually and culturally significant aspects as the theological). Rescher (2000) presents the essential nature of this principle as presented earlier in the literature review for ontology.

The Reality Principle seems to abandon rationalism at the outset. Having already established the potential inconsistencies of rationalism with complexity this should not be disquieting in and of itself, but as it foreshadows a potential bias towards empiricism it is appropriate to acknowledge that bias early in the discussion. In essence, this apparent bias is a consequence of beginning with an ontological perspective as rationalism inherently begins with epistemology. Indeed, since the Reality Principle (among myriad other implications) essentially establishes an ontological priority over epistemology, a segue is necessary to introduce the notions of knowledge. This is in consonance with Rescher (2001) who noted that "an objective domain of impersonally real existence is not a product of but a precondition for empirical inquiry" (pg. 113). This is established through the assertion of a second principle:

O_2: Self-awareness defines a unique existence within reality. (Self-awareness Principle)

With appropriate homage to long-standing discourse related to being, cognition and similar topics, the critical contribution of this principle (in this regard) is the introduction of a unique perspective embedded within the whole. This principle is necessarily tautological in nature and therefore does not constitute a *de facto* Descartian pronouncement (*Cogito, ergo sum*). Descartes spoke to existence as a consequence of cognition while the self-awareness principle speaks to the inevitable consequence of acknowledging self – the delineation of other-than-self (with the acknowledged acceptance of the principle of the excluded middle within this rational context). This is essentially an ontological assertion defining the existence of an epistemology, not an epistemological assertion. From the Reality Principle, both self and other-than-self are contained within reality and this critical disassociation of self from whole-containing-self creates the *foundational duality of cognition* – awareness of distinct existence within a reality otherwise free from such distinction (in this perspective). The unity subsumes all, but the duality is inescapable once created by awareness. For this discussion, self-awareness is only of a unique existence; a corollary notion of awareness will be developed in the epistemology section.

From the introduction of this disassociation it is possible to infer creation and causality – something that was not previously in existence – so that the notions of principal spatial and temporal dimensions are resident within this ontology. However, this is an inference, not an implication – it is possible to posit that awareness precedes self awareness, the converse, or to assert that they are mutually generative and therefore equivalent in some nature. The argument to this point is not a causal argument, but a tautological one that seeks to define and deduce consequences of definitions – it is sufficient that the Self-Awareness Principle and the concomitant foundational duality of cogitation can be asserted to exist. The principal spatio-temporal dimensions will be formally introduced as the argument matures.

A final ontological consideration also foreshadows the epistemological discussion by introducing limitations:

O_3: Everything cannot be contained within less than everything. (Incompleteness Principle)

This is essentially a restatement of the first principle within the conditions of the second principle. Since reality, as that which exists, is irreducible and self awareness defines an existence within reality, reality *in toto* cannot exist within self awareness; hence the disassociation of self from other-than-self must constitute a limited awareness of reality. Using the Kantian formulation of *noumena* being the "thing in itself" (Kant 1901, pg. 248), or "Ding an sich," the *noumena* of other-than-self are not the *noumena* of self and cannot therefore be resident within the self; only *phenomena* can influence the awareness of other-than-self. This is a stronger argument than appealing to the self-imposed bounded nature of self-awareness; it explicitly acknowledges that which is excluded by any bounds that may be established. From a different perspective, this is the negation of the law of identity: that which is not-self is not self, and cannot contain "ding an sich" within self. Further discussions on the fundamental limits of knowledge are found in Rescher (2009b).

An alternate perspective is to assert the primacy of awareness:

O_2: Awareness of other-than-self defines a unique existence within reality. (Awareness Principle)*

This also supports the argument of necessarily limited awareness, but would permit awareness of self preceding awareness of other-than-self (Descartian), or awareness of other-than-self preceding awareness of self. This returns to the illusory nature of a snapshot in a dynamical consideration, reiterating notions of causality that imply a reduction not yet discussed. However, this perspective facilitates discussions of facets of awareness resident within reality prior to awareness itself – but the disassociation is still established from the perspective of self (e.g. self and other-than-self still remain within reality).

Finally, the notion of change introduces temporal structure – that reality includes time and change. Ascribing these properties to reality is tantamount to asserting a positivist quality of something that exists within reality (either derivative of or foundational to other things that exist); likewise to ascribe as derivative of awareness risks asserting a constructivist quality. This question is set aside for this discussion as the essential element for the argument is that time and change are common to awareness, whether its source is foundational to reality or foundational to awareness. A similar discussion holds for the principal spatial dimensions. Hence, the temporal-spatial characteristic asserts that:

O_4: Awareness incorporates the principal temporal and spatial dimensions. (Temporal-Spatial Characteristic)

3.3.2 Epistemology

The result of awareness is generically designated as a cognitive representation of reality (CRR, or "reality as cognitively re-presented", R_{cr}):

E_1: The result of awareness is a cognitive representation of reality. (CRR Principle)

Cognitive is used in the sense of "… pertaining to cognition …" where cognition is "the action or faculty of knowing taken in its widest sense …" (OED 2008). As a defined term, the CRR is quite literally a "re-presention" of reality in the cognitive domain that acknowledges *knowing* refers to something *known*. In this sense, cognition is the process (or processes) that generate the CRR. A critical distinction here is that the boundary between R and A is defined by A given the earlier Self-Awareness Principle; alternatively if the Awareness Principle is accepted then the boundary is mutually generative by interactions within R. In either case, it is possible to delineate from the perspective of the aware or self-aware entity that there exists self (A) and other than self (¬A) such that ($R = A \cup (\neg A)$). The assertion of ($R_{cr} \in A$) is made without additional stipulation such that A may also include elements other than R_{cr}; however, since the CRR is defined as that of which the entity is aware, it is asserted to be wholly contained within the self-aware entity. The domain of awareness D_a – that portion of reality that is accessible to the self-aware entity, or in the case of the aware entity that portion which is mutually generative of awareness – is likewise asserted to wholly contain A, without prejudice to perspectives that consider more detailed aspects of the CRR (note that the domain of awareness is introduced here to facilitate a subsequent discussion on shared domains of awareness). This caveat is relevant as subsequent discussions may touch upon the ideas of explicit vs. tacit knowledge, or conscious vs. unconscious aspects of cognition, wherein the self-aware entity may not explicitly acknowledge some aspect of reality that is accessible to the entity. Under this definition, all that which is accessible to the entity is contained within the domain of awareness; a further delineation of a domain of conscious awareness is a trivial extension to clarify such points if required. Ultimately, the CRR exists because of and reflects the foundational duality of cognition; it exists as part of reality because of the delimitation of a unique existence within reality.

There are some fundamental characteristics of the CRR that can be discussed relating to generation, scope and structure. Fundamentally, the CRR is defined as reflective of reality and is therefore reactive to changes in reality – the first introduction of a causal relationship in this argument. It therefore must have some generative processes which modify the CRR to reflect changes in reality. As the domain of awareness is limited, the CRR is limited and therefore fallible. Also, the CRR is defined relative to the self-aware entity; as described above the domain of awareness includes the cognitive (reality includes self and other-than-self) there are what would be designated rational (internal to the self-aware entity) and empirical (external to the self-aware entity) aspects to cognitive representations. As a limited portion of reality is accessible to awareness, it is possible to adopt a more precise Kantian formulation where *phenomena* are accessible to the domain of awareness and *noumena* are not. In this formulation, the phenomena is that which forms the stimulus for generating the CRR (though the CRR, since it exists, is comprised of noumena and thus is noumenalogically influenced as well).

Acknowledging that the CRR is the product of some generative process does not in itself define the CRR. However, the generative process implies a causal relationship and therefore some type of resulting structure for the CRR – a structure that will be reflective of the generative process. This gives rise to the CRR structure principle:

E_2: Cognitive representations of reality are characterized by a structure reflective of its generative processes. (Structural Principle)

This principle asserts only that the CRR is generated, and consequently has a structure. The CRR structure is inherently a functional representation of the domain of awareness; since there is a causally generated structure, generative processes must extract, classify and organize within the CRR in some form. The notions of a formulation of a CRR that is functionally representative of the domain of awareness simply implies there are entities within the CRR, and there are relationships between those entities. This is not without consequence, as it results in cognitive representations that have systemic characteristics, with systems being used in the sense of "a set or assemblage of things connected, associated, or interdependent, so as to form a complex unity; a whole composed of parts in orderly arrangement according to some scheme or plan" (OED 2008). The notion of abstracting entities within the CRR from the domain of awareness, generalizing aspects across those entities, and instantiating some organization based on this abstraction and generalization results in an assemblage of things (rational and/or empirical in origin) with some relationship, or structure, between those things. However, the presence of systemic characteristics does not necessarily imply a system as the latter requires a comprehensible relationship between the holistic and the particular.

The CRR is therefore broadly asserted to have an associated generative process and a structure, illustrated by the systemically framed set of entities, boundaries and relationships. In order to facilitate discussions relative to the CRR, these may be further assessed relative to their scope and functionality specifically comparing the generative process to worldview (Weltanschauung) and asserting that the structure of the CRR can be conceptually represented as having a *perspective* and *resolution*.

Worldview is a richly debated notion centering on a fundamental orientation reflecting core beliefs, typically discussed as common to a culture or society. In this instance, when examining the notion of a generative process for a CRR, it is apparent that the functionality of that generative process encompasses the broad notion of worldview – that is, how the CRR is generated will be guided by some process that therefore defines the predilections inherent to a particular self-aware entity. In particular, discussions regarding ongoing research (Kovacic 2009, personal communication) explore the idea of "dimensions" of worldviews – the idea that a worldview could be characterized by a set of dimensions, and the interpretation of reality influenced by the degree of the various characteristics. Jones (1972) presents elements of a worldview vector, providing illustrative elements of those characteristics which impact the perception of reality. The work of Bozkurt (2009) further explores the idea that there is a philosophical profile for an individual which impacts the framing of reality. The relevance to the argument at hand is simply that there is a body of research discussing the characteristics of precisely such a generative process as is asserted for the CRR, and notions of worldview and philosophical profiles treat sufficiently similar ideas to frame a shared understanding of this generative process. The generative process can be considered to have characteristics, and the degree to which these characteristics are present will be reflected in the CRR itself.

Having discussed the generative process, the structure of the CRR can be further discussed. First, the degree of abstraction and generalization in the CRR is defined as inversely proportional to the resolution of the CRR. A high degree of abstraction and generalization will result in a distinctly different CRR than a low degree; the organization will necessarily vary accordingly. Inversely proportional is used to employ resolution in a sense analogous to granularity; e.g. a high degree of abstraction and generalization will result in a low degree of detail. Since the resolution will impact the form of organization or structure of the CRR, it characterizes the relationship between the CRR and the domain of awareness (and therefore reality).

Returning to the CRR principle, the fundamental issue is that there is a structure. Generalizing the example above, it is sufficient to say that the structure of the CRR has characteristics that will inherently differ when compared relative to some metric. The metric itself, as a characteristic of a generated structure, will be subject to the generative process; but given a structure there will likewise be characteristics, and given characteristics there will be degrees of those characteristics. When treated in these terms, the notion of resolution (or its inverse, granularity) can be held to be generally applicable to the CRR.

As an illustration of this notion of resolution, consider the typical case of strategic planning for an enterprise. The start point is often taken to be a *Strategic Vision* that expresses a high-level approach to attain a goal. In order to be translated into action, this must be expressed in a fundamentally different form through the expansion via myriad additional details and relationships. This translation echoes a move from the holistic to the particular, and examples abound where the representation of reality at the operational level is different from that at the strategic level. Resolution itself is clearly an essential element to gaining understanding in a limited and fallible CRR; it is related to the essential vagueness of knowledge discussed in (Rescher 2009b).

However, given that there is a granularity to the structure – a metric of scale or scope characteristic of the CRR – it is also necessary to acknowledge the differences inherent between various CRRs. A CRR can be considered to be subject to its location, both spatially (in accordance with the spatial aspect of the Temporal-Spatial Characteristic) and functionally (in accordance with its generative process or worldview). This is expressed via the notion of "perspective," capturing the relative functional or spatial orientation of the self-aware entity with respect to the domain of awareness. As an example, a probabilistic perspective represents one that neglects a dimension to look at behavior over that dimension. Given a process that may have random behaviors over time, a probabilistic perspective neglects time to consider the cumulative behavior as characterized by various statistics. This is an interesting classification as complex behaviors are often plotted on a continuum including deterministic to probabilistic; by this methodology both would represent CRRs with different perspectives. These perspectives are generally intended to provide understanding of the subject events – though not necessarily so; complexity could still be found in either perspective.

Finally, with respect to the Temporal-Spatial Characteristic, the CRR will include change and time. To visit the CRR at any fixed point in time is as illusory as a photograph, but it is a requisite price of analysis in any dynamical arrangement. These "snapshots" of CRRs provide a useful additional concept as CRRs may be considered for multiple temporal instantiations (e.g. the temporal association of the CRR). Examples include:

- Descriptive: The *running CRR*, based on phenomena but presumed to lag reality
- Potential: A projection of what the CRR could become
- Historical: A prior CRR
- Desired: A projection of what the CRR is desired to be

In sum, the CRR is the product of a generative process (broadly aligned with worldview) and subject to that generative process, is characterized by a structure possessing characteristics including analogs to both perspective and resolution.

The Temporal-Spatial Characteristic asserts that the CRR must change as reality changes. The individual, as part of reality, will change and therefore changes in the individual (specifically, the noumena of the individual) will change the CRR itself (since the CRR is wholly contained within the self-aware individual). This can be considered in terms of changes in the generative process for the CRR, or in changes to the structural characteristics of the CRR. In general terms, changes to the structural characteristics may be considered more amenable than changes to generative processes – it is easier to choose, for example, a different perspective and resolution than to change the fundamental parameters of worldview. The critical point is that the CRR can implicitly be influenced, and can be influenced by a matter of cognitive processes. In practice, these cognitive processes may be further subdivided, for example, into conscious and unconscious processes for modifying the CRR.

These elements provide sufficient structure to further shape the definition of a "situation"; specifically that of a construct to frame discussions of complexity relative to reality, or a portion thereof; a self-aware individual (one or more), and the individual's CRR. This is analogous to Sousa-Poza and Correa-Martinez (2005), which frames the discussion using entity (reality or portion thereof); observer (self-aware individual), and solution form (CRR). The asserted equivalence of the CRR with the solution form may seem incongruous, but the differences can be illuminated by considering the work of Maturana (1988) where observation is used to capture not simply gathering data, but the active process of interaction across boundaries – of observer with environment, observer with entity and observer with solution form – capturing the essence of continuous processes relating learning, planning and implementation. This perspective tends to emphasize the Awareness Principle where the environment defines the observer as the observer defines the environment. Since the CRR is simply a label for that which is inherently part of this dynamic process, the apparent differences between the construct are a result of the snapshot illusion – as time progresses, the functionality and interaction of reality, the self-aware entity and the CRR will be equivalent to that of the entity, observer and solution form.[2]

Finally, experience compels recognition that multiple individuals exist within the domain of awareness; so multiple representations of reality may exist across individuals (to be addressed in a subsequent section).

3.3.3 *Knowledge and Understanding*

Knowledge, as used here, is defined as a justified true belief with a reliable context and method for justification (to account for Gettier 1963), or JTB(+). Knowledge is resident within awareness and further refers to the domain of awareness – the

[2] Dr. Sousa-Poza remarked in a 2010 conversation that the solution form may be most analogous to the generative process within PRISM.

individual establishes (is aware of) a JTB(+) for some portion of the domain of awareness. Extending this to more than one individual thus requires establishing mutually intersecting domains of awareness, essentially generating respective representations of reality that are sufficiently congruent to establish a shared context. From this definition, a classification scheme for levels of the CRR follows the lexicon presented earlier for the generalized JTB(+) canons:

- Ideas: Elements of the CRR (not necessarily held to be justified or true)
- Beliefs: Elements of the CRR held to be true
- True Belief: A belief held to be reflective of reality.
- JTB: The assertion that a true belief can be communicated (externally justified).
- Knowledge Claim: Elements of the CRR held to be justified and true with a reliable method and context for justification within the domain of awareness
- Knowledge: Elements of the CRR held to be justified and true with a reliable method and context for justification within intersections of domains of awareness

As presented earlier, within this framework a knowledge claim is internally defined and knowledge is mutually defined.

Recalling the earlier descriptions of understanding from OED (2008) and Dewey (1933), the CRR construct presents understanding as a reflection of the relationships between entities (temporal or spatial) within the CRR – the organization of abstracted and generalized information constituting the means-consequence relationship. This organization is essentially the conceptual context for understanding, varying in degree and completeness (e.g. Nickerson 1985).

3.3.4 Shared Domains of Awareness

The construct for knowledge requires mutually intersecting domains of awareness, essentially generating respective representations of reality that are sufficiently congruent to establish a shared context. An individual's knowledge is some abstracted and generalized representation of reality that is held to be reflective of external reality meeting the JTB(+) condition. This knowledge is asserted to be a function of the degree of the structural characteristics and generative processes of the CRR. In order to build knowledge, there must exist some means to establish across individuals: an individual's CRR; the individual's generative processes for the CRR; the individual's structural characteristics of the CRR, and a reliable method and context to assert it is reflective of reality.

Based on the principles presented earlier, there is but one reality and there is existence within that reality. There can be multiple self-aware individuals within that reality, and there can be change. From the earlier discussion, as awareness is contained within reality, *a change in awareness is a change in reality*. Hence, a change within an individual's awareness is a change in reality that may then be accessible to another self-aware individual if and only if it is phenomenological in nature and accessible to their domain of awareness. Since the domain of awareness is only that which is accessible to awareness, this may also require adaptation of awareness within the domain.

This discussion speaks to bridging across the disassociation from the perspective of the individual: an individual's action is change in reality sourced in the individual's awareness. Within reality there is no bridge required. But, with the acknowledgement of multiple self-aware individuals it is possible to associate a change in reality with a self-aware entity – a phenomena carrying the characteristic of being associated with another unique disassociation. As a result, shared context can be defined by an individual's action: specifically, a change in reality, sourced in one individual's awareness, that is phenomenological in nature and within another individual's domain of awareness. This directly implies that shared context can just happen but can be purposefully constructed and can therefore be located. This brings to mind the notion of action science (Argyris and et al. 1985), and it is directly apparent that individual action of any kind has the potential to build shared context and facilitate mutual knowledge.

Individual actions that are executed with the purpose of establishing across individuals: respective CRR's; respective generative processes for CRRs; respective structural characteristics of CRRs, and reliable method and context asserting they are reflective of reality may be broadly grouped within the general notion of communication. A key point here is that a phenomenon must perceived as being associated with the action of self-aware individual in order to establish communication – the characteristic referred to above conveys a distinction from actions perceived to be otherwise sourced. This line of reasoning returns again to Rescher (2000): "The assumption of a mind-independent reality is essential to the whole of our standard conceptual scheme relating to inquiry and communications" (pg. 97) again recalling the caveat that mind-independent is not mind-exclusive.

With the introduction of action and communication, this begins to approach the notion of the observer as "a cognitive system capable of linguistic behavior" (EA 1998). Varela (1979) asserts at least three main properties characterizing an observer: "(i) capacity for indication: to decide boundaries, to come up with nodes, systems, to have criteria for stability; (ii) capacity for time: to chop a net and start a sequence, to compute through a process, to approximate the stability of a whole; and (iii) capacity for agreement: to externalize, to synchronize with other observers, to re-produce other's distinctions and follow corresponding time patterns" (pg. 65).

The distinctions asserted for the CRR and communications between individuals speaks to the bridge across the disassociation – the interaction of self with other-than-self that mutually defines awareness. In Sousa-Poza and Correa-Martinez (2005) this ultimately resulted in the *solution form*, which encapsulates the bridge across the disassociation. The notions of action discussed above – where phenomena originate with action sourced in the individual's awareness – is reflective of this train of thought; in this instance the bridge is seen more from the perspective of the individual. The two approaches are not incompatible, but provide additional perspective on how various principles within a model can emphasize different aspects within it, ultimately contributing to a richer understanding of the philosophical foundation.

3.3.5 Axiology

The area of axiology introduces a critical component for a philosophical foundation: to discuss the nature of being and knowing is essential, but to choose in a meaningful manner requires a basis for choice and therefore some foundation of value. The axiological component of the philosophical foundation presents the meta-foundations of choice and value – a selection of principles to make explicit the often implicit underpinnings of profound and trivial selections. Specifically, two principles are presented – asserting implications of a particular value and the existence of value itself – that are essential to the development of the objective construct.

An earlier discussion introduced the notion of consistency as a principle, where consistency is a clear proxy for logically sound, semantically complete or various other expressions that imply a clear rational coherence to a basic set of principles:

A_1: *An operation is justifiable if it is internally consistent with its philosophical foundations. (Justification Principle)*

The tautology is present in the equivalence of the term *justification* and *internally consistent* following the definition presented above. This principle is a coherent extension of the law of non-contradiction which is itself an axiomatic statement.

This principle was explicitly applied during the definition of the methodology, which applied the principle internally in the form of constructing internally justified canons referenced to the definition of knowledge, but also when instantiated with respect to a particular philosophical foundation. However, there are direct ramifications when this is placed in the context of the ontological and epistemological principles above; in essence it holds that actions can be justified within the CRR if they are consistent with these foundations. This echoes the debate about the evolution of knowledge; and would for example deem Newtonian physics as knowledge (for earlier times) and would likewise deem relativistic physics as knowledge (for current times). In this instance, new knowledge modified old knowledge, but in some cases it may obviate that knowledge –for example, the notion of æther was accepted and subsequently rejected within essentially the same philosophical foundation (logical positivism). The differentiation serves to clarify the difference between a change in the CRR (e.g. additional content or a revision in perspective and resolution) and a Kuhnian paradigm shift in the underlying philosophical foundations.

Having asserted the Justification Principle, it is necessary to introduce a principle essential to decision-making: that of value. The discussion on decision making presented the necessity of value, and when placed in the context of the Justification Principle the necessity becomes paramount. In order to make a justified decision, there must exist a reason, however it be expressed – a preference or valuation that guides the choice. This valuation must be expressed consistent with the above ontological and epistemological principles. To clearly state an implicit assumption regarding the valuing entity, the decision in question is being made by a self-aware entity (decisions made elsewhere in reality, such as those by another entity, will be discussed at a later point). Since the valuation must express a preference, the preference must be one of things that exist and are therefore in the mind-independent objective reality (Reality Principle). As a result, the preference is, insofar as awareness is concerned, expressed wholly within the entity's awareness. This preference must be expressed relative to a CRR – a preference between projected futures, for example, is a preference between projected CRRs. This is asserted in the Value Principle by defining the notion of a value premise as that premise which establishes the ability to express preference:

A_2: A value premise establishes a preference between CRRs. (Value Principle)

Defining a value premise does not create a value premise; however, it is an essential assumption that a value premise is required in order to make a meaningful decision. Moreover, using the rational methodology for this philosophical foundation, a value premise is essential to justify a decision as it establishes the premises from which conclusions can be drawn.

Since the CRR is limited and fallible (by the Incompleteness Principle), the preference established from the value premise may also be fallible. Taking this to the extreme leads to the perspective of fallabalism, in which case no decision is justified. Sousa-Poza and Correa-Martinez (2005) reject this perspective and asserts the value premise can demand a decision on the part of the self-aware entity, as captured by pragmatic idealism. As an example, the value premise for a self-aware entity may be defining in nature, to the extent of being mutually generative. As an extreme case, life itself – as a self-sustaining biological process – must sustain life or cease to be. A more commonplace example could be in an official of a government making a decision in the course of his or her duties – in the case of the United States government, such officials swear to "support and defend the Constitution of the United States of America" (U.S. Army 2006, pg. 2–2). This can be considered a self- defining value premise; if a decision ceases to conform to this value premise then the official ceases to act as an agent of the government. For these cases, this type of self-defining value premise is defined as a final or ultimate value premise; e.g. one that would be invariant across all CRRs for justifiable decisions (or else it would not meet the criterion of a self-defining value premise). This places the ultimate value premise on a par with the generative process for the CRR – not necessarily a component of the generative process, but one that would impact the structure (perspective and resolution) of the CRR. This analogy parallels the notion of the generative process as worldview, and of that of a paradigm (including essential values) guiding not only perceptions of the world, but methods of investigation and observation.

The ultimate value premise is therefore independent of the structure of a given CRR, yet similar to the arguments that assert a structure for the CRR, it can be asserted that the ultimate value premise will be instantiated in a particular form for a given CRR. This instantiates a particular representation of the value premise concomitant with the representation of reality (e.g. incorporating perspective, resolution and other characteristics of the associated structure). The particular representation of an ultimate value premise within a given CRR is referred to as a penultimate

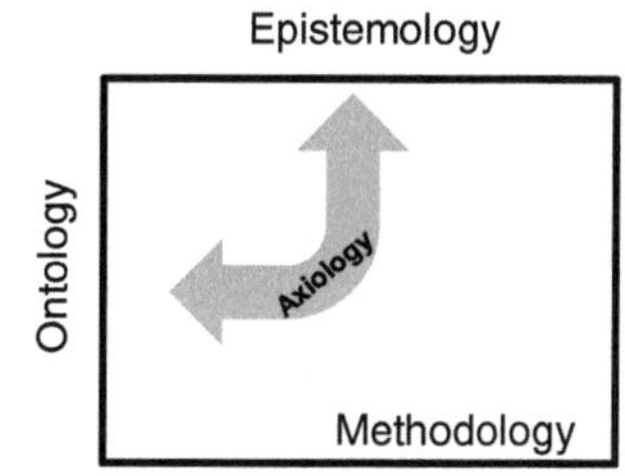

Fig. 3.1 Initial relationship of epistemology, ontology, methodology and axiology

value premise, designating its place as next to final in the establishment of value. This does not exclude the establishment of any number of other, transient value premises; but the notion of an ultimate value premise has a particular significance and its representation within a given CRR will clearly have a central role in decision making, as these other value premises may not have alternative representations in different CRRs.

3.3.6 Methodology

A methodology is "the branch of knowledge that deals with method generally or with the methods of a particular discipline or field of study" (OED 2008). Methods are particular to a circumstance while methodologies guide the development of specific methods. For this model, a methodology must provide guidance for methods that lead from data to understanding and, ultimately, action towards a value premise. Based on the principles provided earlier, this evolution (dynamic and interrelated) will depend on the CRR.

Returning to the Justification Principle, the desired methodology will provide guidance for methods relating ontology to epistemology, consistent with axiology (Fig. 3.1). Consequently, the methodology will be composed of conclusions derived from the associated principles. The task at hand, then, is to determine precisely what epistemological and ontological considerations are to be mapped. In this instance, they must be derived from the considerations relating to the CRR and complex situations. As complexity has been stated to fundamentally deal with understanding, and understanding has been expressed with respect to the CRR, the epistemological components are defined related to understanding; it remains to determine the appropriate ontological components. If the ontological and epistemological components are consistent with the associated principles, then the construct as a whole will be internally consistent.

Understanding, as presented earlier, stated in this manner involves the individual and CRR. By extension, it also involves the broader situation (e.g. the individual situated within reality); within which an individual may assert understanding or the lack thereof. However, given the reality principle and incompleteness principle, a distinct assumption can be asserted with respect to reality itself – namely, that reality may comprehensible, meaning "that which may be grasped" (OED 2008) or

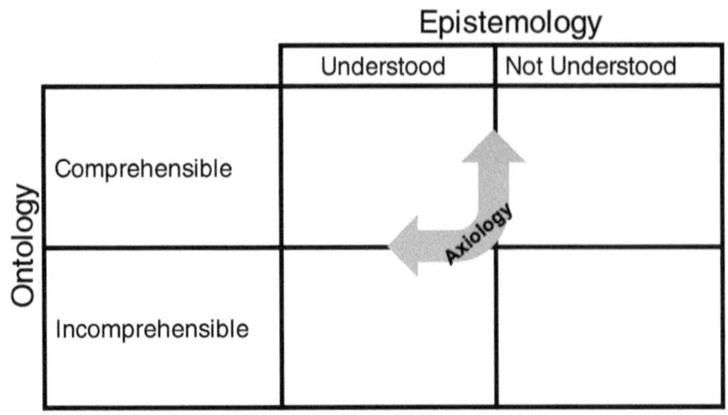

Fig. 3.2 Revised relationship of epistemology, ontology, methodology and axiology

incomprehensible, meaning "that cannot be grasped by the understanding; beyond the reach of intellect or research; unfathomable by the mind" (OED 2008). This implicitly reflects the nature of a complex situation; the situation of comprehensibility (or lack thereof) exists in reality only as a consequence of the existence of a self-aware entity and the associated CRR. Yet, it is clearly an ontological condition; albeit an ontological consideration that is the result of an epistemic consideration. This matures the illustration of the methodology (see Fig. 3.2) to incorporate understanding as the epistemological component and comprehensibility as the ontological component. It also further refines the notion of a situation as having to do with comprehensibility and understanding; in a discussion with Andres Sousa-Poza, a situation was referred to as "a perception of a set of conditions that exist at a point in time … it is an epistemically driven ontological construct". Each of the quadrants, or situations, can be considered individually; however, it should be noted that in making these designations no precedence has been asserted – these are simply epistemological and ontological conditions at this point.

The first situation to discuss is that where the CRR asserts understanding of a comprehensible reality. This situation is bounded by the domain of awareness since the epistemic determinant does not extend past that point (which recalls the inherent fallabilism of the CRR, now extended to the definition of comprehensibility). With respect to complexity, the sources for *error* in knowledge claims can either rest within the CRR (if reality is comprehensible) or within an incomprehensible reality external (from the perspective of the individual) to the CRR. These are rational designations; under the earlier principles the CRR is established as limited and unique to the individual. Similar to the precepts of fallibilism, the CRR is developed from a limited subset of reality (phenomena) and all the CRR of noumena is induced. Kant (1901) summarizes this as:

> Nay, further, this conception is necessary to restrain sensuous intuition within the bounds of phenomena, and thus to limit the objective validity of sensuous cognition; for things in themselves, which lie beyond its province, are called noumena, for the very purpose of indicating that this cognition does not extend its application to all that the understanding thinks. But, after all, the possibility of such noumena is quite incomprehensible, and beyond the sphere of phenomena, all for us is a mere void; that is to say, we possess an understanding

> whose province does *problematically* extend beyond this sphere, but we do not possess an intuition, indeed, not even the conception of a possible intuition, by means of which objects beyond the region of sensibility could be given us, and in reference to which the understanding might be employed *assertorically* (pg. 248).

As it is not possible to prove within the CRR using inductive reasoning, the source of complexity cannot be proven, in general, to reside in the CRR or in reality. Such proofs are necessarily *a posteriori* (again, derivative of the inability to prove comprehensibility in general), and are subject to all of the individual, unique, limited and resolution aspects for the CRR. Consequently, even inferred causality is individual, unique, limited and has an associated resolution. Counterexamples to comprehensibility in reality external to the CRR include any non-causal event inferred in reality (such as beyond the Plank scale, or creation itself) and counterexamples to comprehensibility within the CRR can be found in complexity, paradox and duality (e.g. Goedel's theorem). In a different but related vein, Rescher (2009b) presents a rational argument for the infinity of facts vs. the finitude of truths which reiterates the limits of comprehensibility for a given CRR.

Previous arguments established (under the tenets of the associated principles and derivative conclusions) that, for a given CRR, comprehension can be established *a posteriori*. This walks a fine bridge across the disassociation; for this argument understanding is used as the means-consequence relationship within the CRR, but here there is an assertion, within a CRR, the prior comprehensibility of reality. Comprehension cannot be predicted within a given CRR *a priori* – which would leave no apparent basis for action. However, with the necessary assertion that asserted comprehensibility of a given situation will remain constant within the established resolution a *justifiable* basis for action can be asserted:

M_1: *Within a CRR, establishing comprehension a posteriori defines a justifiable basis for action a priori. (Action Theorem)*

The basis for action is assumed to be movement from the descriptive CRR to a desired CRR; this is used in the sense of direct action towards a value premise as opposed to indirect action such as learning and adaptation including, for example, concepts such as Action Research (Argyris et al. 1985). With the assumption that establishment of comprehensibility can be justifiably projected forward in time, the temporal scale of resolution defines the appropriate transience for action and the spatial scale of resolution defines the appropriate scope for action. This assumption implicitly assumes *status quo* relative to that CRR, an assumption clearly valid for much of human experience, but still an assumption based on the stated ontological and epistemological foundations. To reiterate, assertions of comprehensibility a priori can be *justified within the CRR* but cannot be *proven for reality*. As comprehensibility implicitly refers to phenomena and noumena, comprehensibility can be justified inductively but not proven a priori.

The second quadrant presents a situation where the CRR does not assert understanding, but the situation is comprehensible. Again, this situation is bounded by the domain of awareness. Since the situation is comprehensible, understanding can be achieved and therefore the means-consequence relationship must be attainable. This is an expression of the classical empirical philosophy whereby more information can resolve the lack of understanding. If a CRR does not achieve understanding, a justifiable approach is to attain understanding through learning, through the different application of existing knowledge as well as the acquisition of new knowledge:

M_2: *Within a CRR, lack of understanding justifies learning. (Learning Theorem)*

The third quadrant considers the situation where the CRR does not assert understanding and the situation is incomprehensible. Since comprehensibility is related to a given CRR, if reality is incomprehensible for a given CRR, then the CRR itself must adapt whether through generative processes or characteristics of structure such as perspective and resolution. Though the notion of incomprehensibility is purely a rational assertion, if the situation is incomprehensible, the CRR cannot reflect reality, is in error, and must adapt:

M_3: *Error between understanding and comprehensibility justifies adaption of the CRR. (Adaptation Theorem)*

This requirement to adapt the CRR is significant; once adapted, actions must be justified within the adapted CRR (for example, at a different perspective and resolution). Essentially, all basis for action under the prior CRR are obviated – there is no assurance that value premises valid within the earlier CRR will be valid within the adapted CRR. Here, the potential of an ultimate value premise may permit the instantiation of a new penultimate value premise.

The final quadrant presents the situation where the CRR asserts understanding of an incomprehensible situation. In this instance, the source of complexity is clear – the error rests with asserting a false understanding. The justifiable action in this situation is to assess the comprehensibility of reality – whether challenging assumptions, models, conclusions, perspective, or resolution. No principle is asserted for this situation as the need to assert comprehensibility by establishing understanding has already occurred; it is simply in error. This implicitly recognizes the limited and fallible nature of the CRR.

From the earlier discussion, it is not possible to determine the comprehensibility of reality *a priori*. Experience compels acknowledgement that comprehension can be safely asserted within much of the scope of human experience, but this remains an assertion based on the assumption of status quo and subject to the constraints of the CRR. As in the last situation considered, it is typical to operate on the basis of understanding – but in cases of complexity, it is essential to abstract not only relative to understanding but relative to the generation of understanding. As a result, there is a consequence of using the natural tendency to consider the ontological-epistemological relationship matrix as one composed of quadrants; this rational distinction created to facilitate understanding presents these distinctions as having an unwarranted clarity. To remove this illusion of clarity, each of these quadrants is referred to as a *situation*, the totality of them representing *complex situations*. With this distinction, the equivalent relevance of these fundamental situations is asserted as the Situations Theorem:

M_4: *Absent additional information, each fundamental situation related to comprehensibility and understanding is equally relevant at any given time. (Situations Theorum)*

Table 3.1 Matrix of situations

Reality is: \ CRR asserts:	Not understood	Understood
Comprehensible	High complexity Justifiable option: Learn	Low complexity Justifiable option: Action within justified CRR
Incomprehensible	High complexity Justifiable option: Adapt CRR	High complexity Justifiable option: Assess CRR

These four situations related to understanding and comprehensibility including an examination of complexity and the dominant appropriate options for the individual are presented in Table 3.1. In situations where understanding and comprehension are inconsistent, the appropriate option is to learn – to build knowledge and context for understanding enabling the establishment of a means-consequence relationship. In any situation of high complexity the appropriate option is to adapt the CRR either through a change in resolution or adapting the parameters of abstraction, generalization and organization (worldview). Finally, consistent with the Justification Principle, the appropriate option is to act within the appropriate resolution whenever comprehensibility is established for that perspective, resolution and situation.

Per earlier principles, it is not possible for the individual to prove the currently applicable situation this matrix – absent additional information, each situation is equally relevant at any given time and appropriate options must be executed simultaneously and continuously. Therefore, the justifiable options for an individual at any given time are to:

- Act when within justifiable perspective and resolution;
- Learn when understanding is needed;
- Adapt CRR when complexity is present, and
- Assess assertions of comprehensibility.

These can be summarized as assessment, action, learning and adaptation (noting the risk in providing equally relevant options in any order). These options will be subject to several challenges in practice – the justifiable perspective and resolution for action may be incompatible with the individual's value premise, learning presents its own suite of challenges, and adapting the CRR through generative processes or structural characteristics may require extended effort and time. However, adapting resolution is common – it is typical, for example, to consider strategic, operational, and tactical levels of detail – and so provides an amenable option for attaining understanding consistent with comprehensibility.

In all of these situations, these justifiable actions are those that are deemed to have primacy of purpose – for example, adaptation and assessment will certainly require learning. These justifiable actions provide a characterization of the situation and are not exclusive. Learning, as a justifiable action, is *learning to gain understanding*, and does not restrict learning to operations within that situation alone (which cannot be clearly separated). In a more general form, the above matrix would guide the individual to learn and adapt when paradox or conflict is seen in conclusions, and act and assess when not. Similarly, when complexity is likely, additional emphasis would be given to learning, assessment and adaptation. Recalling the potential for a penultimate value premise (the instantiation of the invariant ultimate value premise for a particular CRR), each of these justifiable actions represent an antepenultimate value premise – a clear assignation of value for that particular situation – that asserts value based solely on these theorems, derivative of the Justification Principle. This antepenultimate value premise is predicated on the existence of a penultimate value premise – there must exist a reason, instantiated within the given CRR, to act, assess, adapt and learn – as an essential consequence of the Value Principle. In the context of the Situations Methodology, it is also interesting to compare this to the notion from Keeney (1992) where value-focused thinking leads to decision-making opportunities.

An immediate (and inaccurate) comparison can be made to various temporally oriented decision cycles, such as the *Observe-Orient-Decide-Act* (OODA) loop (Boyd 1986, 1987). A decision cycle is fundamentally a temporal process whereas the situations methodology represents justifiable situations at a point in time. This, however, does frame a problem for reduction to practice: if all situations are equally relevant absent additional information, what is the guidance for the practitioner? This issue must be guided by the Justifications Principle, which will allow operations in any one of the situations when canons for truth, justification, method and context are fulfilled. In particular, the method must be consistent with the situations methodology. Since this is a continuous situation, the probability of precisely equal justification for all four situations is theoretically zero (though they may be approximately equal within a given resolution). In the theoretical case, there will always exist a preference to operate in one of the four situations; and in the practical case a decision to assert primacy of any one of them is justifiable. Also, experience compels that there is often sufficient additional information to assert primacy of one of the four quadrants. The asserted primacy of a given situation may leverage, for example, the initial consideration of the nature of the situation at hand – if it is a situation where one may rationally assert comprehensibility given the likely CRR (for example, a well-framed optimization problem) then one would be guided to emphasize learning and action (the options associated with a comprehensible domain of awareness). Since it is necessary to gain (or at least confirm) understanding to justify action, primacy would be granted towards learning as the initial situation- though, per the Situations Theorem, this learning should still contextually account for assessment, adaptation and action. Alternatively, the situation at hand may be one where there is a higher probability of an incomprehensible domain of awareness given the likely CRR – namely, any of the conditions known to give rise to complexity. In those instances, primacy of operations would be granted towards assessment and adaptation.

Having therefore asserted that it will be practical to assert primacy of a quadrant within the situations methodology, and therefore asserted guidance for selection of a justifiably reliable method, the issue of a justifiably reliable context can be formally considered. At a minimum, the context must be sufficient to satisfy the requirements of communication established earlier. Similar arguments to the one that determined the primacy of a situation can also be established to assert the relevance of the other situations. Disregarding the other potential situations leads to a necessarily incomplete representation, leading to the Contextual Corollary:

M_5: *Justifiable operations within a given situation must include all other situations as relevant context. (Contextual Corollary)*

The Contextual Corollary is therefore a necessary condition for cognitive operations within a complex situation. There is no assertion of sufficiency as other factors must inherently be considered – for example, the acquisition of knowledge, consideration of value, etc. must still be present.

The exploration of these theorems, consistent with principles of pragmatic idealism, illustrate the potential of a dialog within a philosophical framework – the clarity of expression for the Structure Principle enables the discussion of justifiable resolution for action, and frames the situations methodology.

3.3.7 Reconciliation with Canons

This section instantiates canons of research for this particular philosophical foundation based on the generalized JTB(+) canons presented in (Brewer and Sousa-Poza 2009) and assesses the results with respect to the instantiated canons. The philosophical foundation established to this point is summarized as a Pragmatic-Rational-Idealist-Situational Model founded on the premises of a mind-independent objective reality (Rescher 2000) containing self-aware individuals with necessarily fallible representations of reality. As such, a belief is one or more propositions (themselves constructs within the individual's representation of reality) that constitute a set of conclusions, from phenomenological sources, that generate understanding – an assertion of comprehensibility. Therefore, an internally consistent canon for truth is demonstration of the internally consistent coherence of the asserted truth with a set of conclusions through an assertion of comprehensibility based on phenomena. This implicitly carries the constraints and limitations of the individual's representation of reality – fallible, limited, and subject to the resolution, generation and structure of the representation of reality – but carries with it a specific tie to the objective reality. Given a self aware individual as a component of an objective reality, the interactions of that individual with reality are similarly a component of the objective reality and create phenomena subject to incidence on other individual's domains of awareness. Though this definition of a truth canon is internal to an individual's representation of reality, it is a definition permitting the extension of the asserted truth to others.

The generalized canon of justification asserts that research must establish that truth external to the individual. This introduces the notion of an external representation of reality (one or more), and requires the individual to assert the justification as accessible to another representation of reality. The elements of the representation of reality asserting truth must be representable through phenomenological action, consistent with the criteria for communication. Specifically, this must include the ability to communicate the philosophical foundations, the set of propositions against which the truth establishes coherence, and the phenomenological sources that generate those conclusions (with the concomitant assertion of comprehensibility). As a practical example, an individual may hold some data set to constitute justification for a truth – the data set represents the phenomenological sources, and the analysis of that data set represents a set of propositions against which the truth is justified. The individual then must believe the sources, data, analysis and resulting propositions can be established in a shared domain of awareness with some other individual(s). This does not guarantee that another individual or individual(s) will accept this truth, but a necessary condition for justification is that the individual asserting the JTB(+) believes this to be possible.

The remaining two generalized canons, method and context, being the ways and means of justification, can be taken to represent a process and substantive perspective on the actual establishment of a shared domain of awareness. A shared domain of awareness must incorporate two or more representations of reality; by direct implication there must be phenomenological action sourced in the respective representations of reality that is within the respective domains of awareness – e.g. there must be communication involving, and active effort for, both. Returning to the critical point of communication, if phenomena are not perceived as sourced in another representation of reality (purposively or neglectfully) then the essential characteristic of sourcing within a self-aware individual is not present and the ability to purposefully construct a shared domain does not exist. The definition of shared ways and means within the shared domain of awareness requires multiple representations of reality to participate – quite common in practice as numerous references on research methodology are available, providing specific ways and means to establish a shared representation of reality within specific philosophical foundations.

Under the philosophical foundations in question (pragmatic idealism and associated derivative principles), the generalized canons of method and context are asserted to be a shared ways and means by which a shared domain of awareness is constructed. A reliable method and context must be one which is held to be justified across multiple representations of reality. The method and context itself becomes a JTB(+) requiring a method and context; ultimately this infinite regress must terminate in the philosophical foundations which here require acknowledgement of an (limited, fallible and unique) assertion of comprehensibility based on phenomenological sources. As these sources are held to be derived from an objective reality, they are potentially accessible to multiple representations of reality and therefore permit this assertion to be shared. Those elements of reality accessible and accessed by relevant representations of reality then constitute the means by which the shared domain of awareness is constructed (though this can clearly be constrained to those relevant to the JTB(+) assertion at hand). These elements are both objectively and mutually defined; hence the shared domain is immediately seen to be dynamic in nature – it is defined by reality and by the relevant representations of reality. Context for justification will necessarily be dynamic as well. The ways by which the shared domain of awareness is constructed are then those elements of communication which enable the establishment of the domain; a method ultimately speaks to enabling communication via the instantiation of a body of concepts in a manner accepted by multiple individuals.

The philosophical foundation derived above can now be for each of these instantiated canons will be considered in turn. In essence, this serves as a final confirmation of internal consistency – the consideration of a philosophical foundation against canons instantiated for that philosophical foundation – while framing the evaluation against the specific definition of knowledge as justified true belief with a reliable method and context for justification. Truth is asserted through acceptance of ontological, epistemological, and axiological principles. Justification is asserted through internal consistency of those principles. The foundational principles are explicitly presented and founded on a well-established philosophical basis. The propositions (theorems) are rationally derived from the principles, and conclusions and sources are justified relative to the specified domain (a general consideration of philosophical foundations for complex situations). Reliability of method is provided by adhering to accepted practices for rational derivations. The principles are semantically complete (e.g. expressed as tautologies) and the theorems are sound (e.g. derived from the principles). The notable exception is the introduction of an essential characteristic; however, this is such a fundamental characteristic that the likelihood of rejection in the shared domain is held to be trivial. Reliability of context is provided by basing principles and conclusions on established definitions common to the literature, and on explicit rational derivations where extended beyond the literature. In total, the above canons are held to be satisfied in that a shared domain of awareness can be constructed via explicit rational derivations based on principles available in extant literature. The philosophical foundation is therefore a valid JTB(+) knowledge claim.

3.3.8 Methods and Methodology

Consider the value premise of a methodology, which rests in guiding the selection of methods. In the case of the situations methodology, areas for future research include the classification of extant methods as well as the generation of new methods. The former allows consideration (as seen in the illustrative comparisons) of how existing methods may be adapted to operate under the PRISM. For comprehensible situations, there are numerous existing methods that either frame learning (e.g. Action Science) or frame learning to gain understanding and transition to action (e.g. Hard Systems Engineering, Risk Analysis, Decision Theory, and analytical techniques in general). The learning organization approach also tends to emphasize the comprehensible. Methods devoted specifically towards action include such well-known approaches as Project Management, Configuration Control, and Process Control; for highly comprehensible and well understood situations this would include Lean Six Sigma or similar approaches. As the situations transition to high

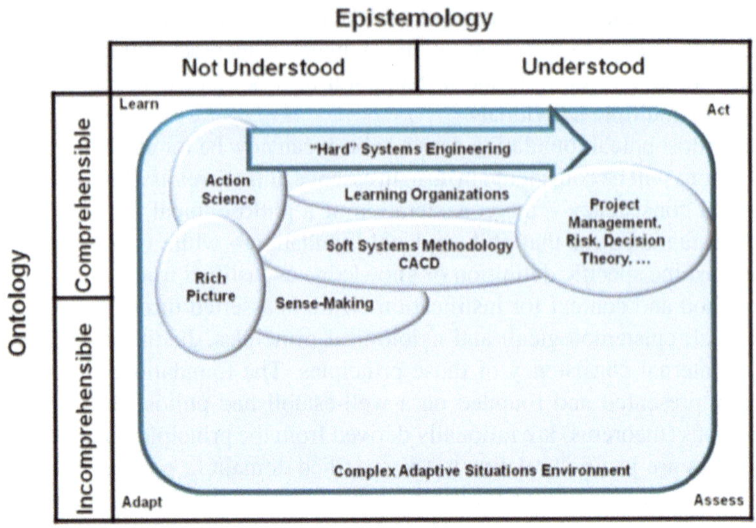

Fig. 3.3 Illustration of methods within the situations methodology

complexity, methods such as Soft Systems Methodology come into play – but it should be noted that this is somewhat sequential in practice, and SSM will transition from learn, to assess, to adapt, and possibly return before transitioning to action. This can be seen as a necessary shift of primacy given the situation, but can also be a by-product of an overly simplistic treatment of a methodology applied as a sequential method – Checkland (1999a) cautions against treating the methodology as a method. The technique of developing a rich picture is not restricted to SSM, and can be seen as either a learning approach or an adaptation approach (depending on the intent for the development of same). Sense-making, as described earlier, may be particularly well suited as a method for adaptation and assessment – but it represents a methodology in and of itself, implying that there may be a finer level of detail in leveraging this approach. Still, the general concepts of communications to enable individual assessment and adaptation are appropriate in any one of the situations where either there is a lack of understanding or of comprehensibility. Finally, there are emerging environments such as the Complex Adaptive Situations Environment at Old Dominion University (Sousa-Poza 2009, personal communication) which are expressly founded on the philosophical foundations of Pragmatic Idealism and therefore have a high degree of applicability across all situations. Figure 3.3 presents a summary of these methods mapped to the situations methodology, illustrating the general area of applicability for each. Hard systems engineering is represented uniquely by an arrow as it represents a well-structured approach for dealing with extremely complicated problems in a manner that allows specification and attainment of well-defined requirements. However, in those cases where requirements are ill-defined, or the problems become complex, hard systems engineering can be severely challenged.

3.4 Conclusion

An explicit philosophical foundation extends our perspective; we can discourse across disciplinary boundaries when they are explicit. As an example, some inspection of the situational methodology reveals the presence of a more basic thread: the Baconian scientific method itself is a linear maturation within the model. It follows the path from assessment (determining that current theory does not explain observation), to adaptation (developing new theories), to learning (developing the justification) and finally to action. As uncertainty is inherent in the basis for science, it is interesting how it became to be affiliated with an inconsistent philosophical foundation of logical positiviem (Rosenberg 2005). Another interesting aspect is that a Kuhnian paradigm shift can take place within PRISM – because it is a foundation that requires the ability to adapt the CRR based on its inherent fallibility. This implies that designating a paradigm as, itself, a philosophical foundation is essentially conflating distinct concepts – a conclusion which led to the shift in designation of PRISM as a general "model".

The formality of the PRISM derivation may seem excessive, but it is in this formality that existing concepts can be structured to provide new insights – the purposeful construction of a methodology, woven with axiology to form a consistent basis for the meta-discussion on complexity. From this basis we can extend the meta-discussion into the cognitive, systems, or other related domains relieved of fear of paradox or dichotomy by this prime directive of cognitive rationality – a rigorous commitment to internal consistency, from the foundation up.

The Pragmatic – Rational – Idealist – Situational Model

Ontology: Pragmatic

O1: *Reality is (that which exists). (Reality Principle)*

O2: *Self-awareness defines a unique existence within reality. (Self-awareness Principle)*

O2:* *Awareness of other-than-self defines a unique existence within reality. (Awareness Principle)*

O3: *Everything cannot be contained within less than everything. (Incompleteness Principle)*

O4: *Awareness incorporates the principal temporal and spatial dimensions. (Temporal-Spatial Characteristic)*

Epistemology: Rational

E1: *The result of awareness is a cognitive representation of reality. (CRR Principle)*

E2: *Cognitive representations of reality are characterized by a structure reflective of its generative processes. (Structure Principle)*

(continued)

(continued)

Axiology: Idealist

A1: An operation is justifiable if it is internally consistent with philosophical foundations. (Justification Principle)

A2: A value premise establishes a preference between CRRs. (Value Principle)

Methodology: Situational

M1: Within a CRR, establishing comprehension a posteriori defines a justifiable basis for action a priori. (Action Theorem)

M2: Within a CRR, lack of understanding justifies learning. (Learning Theorem)

M3: Error between understanding and comprehensibility (complexity) justifies adaption of the CRR. (Adaptation Theorem)

M4: Absent additional information, each fundamental situation related to comprehensibility and understanding is equally relevant at any given time. (Situations Theorem)

M5: Justifiable operations within a given situation must include all other situations as relevant context. (Contextual Corollary)

References

Aliseda A (2004) Logics in scientific discovery. Found Sci 9:339–363

Argyris C, Putnam R, Smith DM (1985) Action science: concepts, methods, and skills for research and intervention. Jossey-Bass, San Fransisco

Aristotle (1994) Metaphysics. (trans: Ross WD). Original work circa 350 BCE. Retrieved 31 Aug 2008 from http://classics.mit.edu/Aristotle/metaphysics.html

Aschenbrenner K (1971) The concepts of value: foundations of value theory. Humanities Press, New York

Azzouni J (2004) Theory, observation and scientific realism. Br J Philos Sci 55:371–392

Boyd JR (1986) Patterns of conflict. Briefing. Retrieved 6 Dec 2009 from http://www.d-n-i.net/boyd/pdf/c&c.pdf

Boyd JR (1987) Organic design for command and control. Briefing. Retrieved 6 Dec 2009 from http://www.d-n-i.net/boyd/pdf/poc.pdf

Bozkurt I (2009) Developing a philosophical profile of the individual for complex problem-solving through agent-based modeling. PhD dissertation, Old Dominion University, Norfolk

Brewer V (2010) A decision making construct for complex situations. PhD dissertation, Old Dominion University, Norfolk, May 2010

Brewer V, Sousa-Poza A (2009) Generalized research canons for JTB(+) knowledge. n: Proceedings of the international symposium on peer reviewing: ISPR 2009, June 2009, London

Checkland P (1999) Soft systems methodology—a 30-year retrospective. Wiley, New York

Cilliers FP (1998) Complexity & postmodernism: understanding complex systems. Routledge, New York

Dewey J (1933) How we think: a restatement of the relation of reflective thinking to the educative process. D.C. Heath & Co., Boston

EA (1998) Encyclopedia Autopoetica, online edition. In: Randall W (ed) Retrieved Aug 30–Sept 5 2008 from www.cybsoc.org/EA.html

Feyerabend P (1981) Realism, rationalism and scientific method, vol 1, Philosophical papers. Cambridge, New York

Georgiou I (2000) The ontological constitution of bounding-judging in the phenomenological epistemology of von Bertalanffy's general system theory. Syst Pract Act Res 13(3):391–424

Gettier E (1963) Is justified true belief knowledge? Analysis 23:121–123

Guba E, Lincoln Y (1989) Fourth generation evaluation. Sage, London

Hart SL (1971) Axiology – theory of values. Philos Phenomenol Res 32(1):29–41

Heylighen F, Cilliers P, Gershenson C (2007) Complexity and philosophy. Retrieved 6 Dec 2009 from http://uk.arxiv.org/ftp/cs/papers/0604/0604072.pdf

Jessup B (1949) On value. In: Lepley R (ed) Value; a cooperative inquiry. Columbia University Press, New York

Jones W (1972) World views: their nature and their function. Curr Anthropol 13(1):79–80, 81–109

Kant I (1901) Critique of pure reason. 2nd edition. (trans: Meilkejohn J). Retrieved from http://books.google.com/books (Original work published 1787)

Keating C, Sousa-Poza A, Kovacic S (2005) Complex system transformation: a system of systems engineering (SoSE) perspective. In: 26th ASEM national conference proceedings, Virginia Beach, Virginia, USA, pp 200–207

Keeney R (1992) Value focused thinking: a path to creative decision-making. Harvard College, Boston

Kieseppä IA (2000) Rationalism, naturalism, and methodological principles. Erkenntnis 53(3):337–352

Lauden L (1986) Methodology's prospects. Proc Bienn Meet Philos Sci Assoc 1986(2):347–354

Lehmann L (2004) Paradigm development in systematic theology. Master's thesis, University of South Africa, November 2004

Maturana H (1988) Ontology of observing: the biological foundations of self consciousness and the physical domain of existence. In: Donaldson R (ed) Texts in cybernetic theory: an in-depth exploration of the thought of Humberto Maturana, William T. Powers, and Ernst von Glasersfeld. American Society for Cybernetics, Felton, CA, [conference workbook]

Miller A (2005) Realism. In: Zalta EN (ed) The Stanford encyclopedia of philosophy (Fall 2008 Edition). Retrieved 15 Mar 2009 from http://plato.stanford.edu/entries/realism/

Mingers J (2003) A classification of the philosophical assumptions of management science methods. J Oper Res Soc 54(6):559–570

Nickerson R (1985) Understanding understanding. Am J Educ 93(2):201–239

OED (2008, 2009) Oxford English dictionary, online edition. Retrieved 30 Aug–5 Sept 2008 and 18–20 Mar 2009, from http://www.oed.com

Olssen M (1996) Radical constructivism and its failings: anti-realism and individualism. Br J Educ 44(4):275–295

Plato (1892) The dialogues of Plato translated into english with analyses and introductions. By B Jowett, M.A. in five volumes. 3rd edition. Oxford University Press, 1892. Original Work c. 340 BCE. Retrieved 1 Sept 2008 from http://oll.libertyfund.org/title/766/93700

Popper K (1934) The logic of scientific discovery. Basic Books, New York

Rescher N (1994) Review: précis of a system of pragmatic idealism. Philos Phenomenol Res 54(2):377–390

Rescher N (1998) Complexity: a philosophical overview. Transaction, New Brunswick

Rescher N (2000) Process philosophy: a survey of basic issues. University of Pittsburg Press, Pittsburgh

Rescher N (2001) Cognitive pragmatism: the theory of knowledge in pragmatic perspective. University of Pittsburg Press, Pittsburgh

Rescher N (2009a) Aporetics: rational deliberation in the face of inconsistency. University of Pittsburg Press, Pittsburgh

Rescher N (2009b) Unknowability: an inquiry into the limits of knowledge. Rowman & Littlefield, Lanham

Rosenberg A (2005) The philosophy of science: a contemporary introduction. Routledge, New York

Sinn S (1998) A comparison of interactive planning and soft systems methodology: enhancing the complementarist position. Syst Pract Act Res 11(4):435–453

Sousa-Poza A, Correa-Martinez Y (2005) Pragmatic idealism as the basis for understanding complex domains: the trinity and SOSE. In: 2005 IEEE international conference on systems, man and cybernetics, vol 3, Waikoloa, Hawaii, USA, pp 2744–2750

Stacey R (2000) Complexity and management: fad or radical challenge to systems thinking? Routledge, New York

United States Army (2006) Field manual 6-22: army leadership competent, confident, and agile. Department of the Army, Washington, DC

Van Uden J, Richardson K, Cilliers P (2001) Postmodernism revisited? Complexity science and the study of organizations. J Crit Postmod Organ Sci 1(3)

Vandenberghe F (1999) "The real is relational": an epistemological analysis of Pierre Bourdieu's generative structuralism. Sociol Theory 17(1):32–67

Varela F (1979) Principles of biological autonomy. Elsevier (North-Holland), New York

Chapter 4
Understanding and Complex Situations

Jose J. Padilla

Abstract This paper presents a theoretical construct for assessing wicked problems as they occur in complex situations and defining problem context. The construct of understanding presents a way for assessing a problem as perceived by an observer. It is suggested that the paradigm of sole full analysis in a complex situation is not feasible when these two conditions are present and pervasive. It is suggested that a synoptic perspective (high level perspective) may be more useful to establish not only what is perceived as reducible and transient within a situation, but also to assess the individual that perceives the situation. To elaborate on the construct, fuzzy logic is suggested as a way to quantify the perception of irreducible and transient variables.

Keywords Understanding • Complex Situations • Problem Situations

4.1 Introduction

Note: As this chapter is an elaboration of a conference paper (MECS 2007), the author has left the original material for the reader to see the original context of the document. Changes were made to the title, portion of the abstract, Sects. 2.6 *and* 2.7 *and new references were added. The changes made are to reflect current research.*

The term "wicked problem", originally credited to Horst Rittel and Melvin Webber, "are a class of social system problems which are ill-formulated, where the conflicting values, and where the ramifications in the whole system are thoroughly confusing" (Churchman 1967). Churchman states that the term "wicked", makes

J.J. Padilla (✉)
Virginia Modeling, Analysis and Simulation Center, Old Dominion University, 1030 University Blvd, Suffolk, VA 23435, USA
e-mail: jpadi002@odu.edu

S.F. Kovacic and A. Sousa-Poza (eds.), *Managing and Engineering in Complex Situations*, Topics in Safety, Risk, Reliability and Quality 21, DOI 10.1007/978-94-007-5515-4_4,

inference to the illicit nature of these problems, describing the "mischievous and evil quality of these problems, where proposed "solutions" often turn out to be worse than the symptoms". Rittel and Webber present them as a type of problems that are complex to formulate and frame and "as distinguishable from problems in the natural sciences which are definable and separable and have solutions that are findable, the problems of governmental planning – especially social and policy planning – are ill-defined" (Rittel and Webber 1973). Organizationally, wicked problems exist where not only technical complexity may be high, but also behavioral complexity. In these cases different groups of decision makers hold different assumptions, values and beliefs that may be in opposition with each other (King and Rittel in Kovacic et al. 2006).

Rittel and Webber proposed ten characteristics to distinguish this type of problems.[1] Some of the most relevant are: (1) *there is no definitive formulation of a wicked problem*: For an individual to understand the problem it would depend upon his/her idea of solving it and possibly develop an inventory of possible solutions ahead of time. The formulation of a wicked problem is the problem. (2) *Wicked problems have no stopping rule*: "The process of solving the problem is identical with the process of understanding its nature" (Rittel and Webber 1973). (3) *Solutions to wicked problems are not true-false, but good-or-bad*: There is no unique or optimal solution to a problem, but several of them where one could be better than the others. This is consistent to what H. Simon said: "as humans, we make decisions or take actions – in most cases – based on incomplete information to reach satisfactory decisions" (Waldrop 1987). (4) *There is no immediate and no ultimate test or solution to a wicked problem*: Any solution implemented, will generate waves of consequences over an extended and unbounded period of time.

Wicked problems and the characteristics that they manifest generate great difficulties for both research and practice. The consequence of the characteristics is that common methods and approaches become fallible due to their assumption of an objective, stable, linear, terminal problem domain. The non-monotonic synopsis serves as a starting point from which a better understanding of the nature of these problems can be understood within the construct of a complex situation.

4.2 Complex Situations

According to (Kovacic et al. 2006) Complex Situations are those from which wicked problems are derived that will often include both reducible and irreducible components. They not only have a complex technical component, but also a complex social component and are the reflection of the behavior of a system of systems. The sheer size of most system of systems being studied results in convoluted structures, and the identification of problems and the capability of solving them become complex

[1] Refer to Rittel and Webber (1973) for the complete list and explanation of these characteristics.

Fig. 4.1 Complex situations

in nature. W. Brian Arthur (2000) says that when problems are too complicated to afford solutions or when they are not well-defined, agents face not a problem, but a situation and that to deal with the situation they must frame the problem, and that framing in many ways is the most important part of the decision process.

Kovacic, Sousa-Poza and Keating propose that these situations are formed through the interaction of three major elements: a domain or entity, solution form, and observer. The *entity* is what we are trying to understand, *observer* relates to the perception capability of the agent rating the situation, and *solution form* relates to the form of the expected outcome (Kovacic et al. 2006). Figure 4.1 shows the concept.

The *entity* is usually associated with the number of components, relationships in a systemic perspective. This generates a level of complexity itself given that the addition of components increases the number of possible relationships exponentially generating an intractable problem. The inclusion of the *observer* becomes crucial to deal with situations that are considered complex, in other words, complex situations cannot be considered without the observer's perspective. In this case, the perspective of the observer may increase or decrease the complexity of the situation. The *solution form*, as it was mentioned above, relates to the granularity required by the observer to develop a solution. A situation then is created by the interaction between observer, entity and solution. The observer when trying to deal with a situation, he/she elaborates a solution that works according to his/her won beliefs and values what he/ she has been able to assimilate via different means.

Note that the word used is *assimilate* instead of *analyze*. This differentiation is proposed to highlight that through analysis, much of the information may be lost, given that it cannot be put back together through a process of synthesis because of the wicked problem no stopping rule. Analysis will generate more analysis and so on. Kovacic et al. (2006) go on to elaborate that "typical complex situations from which wicked problems are derived, will often include reducible components and irreducible ones such as politics and other contextual factors creating a situation where the solution may require a high level of granularity which is incongruent with the irreducible components resulting in low levels of understanding" (Kovacic et al. 2006) This prompts the need for addressing those irreducible components in a non-analytical way.

4.3 Complex Situations Redefined

To redefine what a complex situation is will not add much to the previous discourse. However, this is needed to introduce as mentioned, a way to deal with them that is consistent with the definition of wicked problems and complex situation. This approach introduces one assumption: The entity is complex for every single observer ranging from epistemic to aleatory uncertainty.[2] However, a particular observer may be able to deal better with a given entity as he/she is better suitable for dealing with epistemic and aleatory uncertainties in a given situation.

Based on this assumption, the model of entity, observer and solution form is presented in terms of the reducibility of the situation, the transience of the situation, and the profile of the observer that reflects on how he/she understand the situation. What this new representation does is that does not only reflect the entity of the situation in a temporal and spatial scale, but also associates the solution form to the better capability of understanding of the observer a personal profile. In other words, an observer with certain traits may be better able to deal with a complex situation given its reducibility and transience.

4.3.1 Situational Reducibility/Irreducibility

Reducibility can be defined as the maximum abstraction or maximum analysis obtainable of a situation without changing the nature of the situation at a level that is of use for the observer. Reducibility can be seen under two lights, system decomposability which all system theory fall under and construct reducibility such as the one used to test hypotheses and correlation among variables.

In the former case, reducibility can be seen as the concept of decomposability proposed by Bar Yam (2005). In this, the idea is to identify all possible elements and relationships among elements. Irreducibility is reached when the capability to analyze and synthesize a system based on its components, is not possible (Bar-Yam 2005). Reducibility then becomes an issue of granularity, until what point we can analyze a system without loosing the behavior of the whole? How do we account for what's lost? What's the effect of that which is lost? These are questions that academics and practitioners ask on regular basis and there is no easy answer. But the focus is not only to see if something can be reduced or not, but what to do when is non-reducible. Behe (1996) defines irreducible complexity as "a single system which is composed of several interacting parts that contribute to the basic function

[2] What can be known is associated in this case with epistemic uncertainty, and what cannot be known (and perhaps can be inferred) with aleatory uncertainty. In the reliability literature epistemic uncertainty and aleatory uncertainty are used within this context. The former can be overcome by adding more information while the latter cannot and different approaches may be needed to deal with it (Hora 1996; Durga et al. 2006).

Fig. 4.2 Reducibility/irreducibility in a systemic context

and where removal of any one of the parts causes the system to effectively cease functioning". Figure 4.2 presents the concept of reducibility/irreducibility. If the situation is seen as a system, the figure shows: porous boundaries, irregular granularity. The porous boundaries presents the possibility of new parameters appearing convoluting the existing situation. Finally, the irregular granularity presents that certain elements cannot be reduced without losing its overall behavior.

In a construct sense, reducibility can be presented as the capability of a construct to be explained by conforming variables. In this sense, reducibility can be seen as variables that can be measured directly and crisply, whereas irreducibility as variables that are measured indirectly (proxies) or variables that cannot be measured by regular means. What is important is to define the problem and the identification of the conforming variables is a matter of completeness and their measurability. What is the behavior of the element/variable and the effect it has on the rest of the system/situation? The same applies for relationships; we have identified a relevant relationship between two elements, but how much is known about the relationship? What kind of relationship is it? What is the effect of this relationship on the involved elements of the whole? In any given case, action is needed, a decision is required and that is when the level of uncertainty is assessed to proceed.

4.3.2 Situational Transience/Intransience

Transient behavior is a not a clearly defined term (Bell 1989). Bell defines it as time dependent system behavior that differs from long-term or steady-state behavior. For a system to be steady-state the probability that the system state variables have certain values is independent of time, whereas a non-stationary state, the probabilities of finding one or more state variables at certain values is a function of the time elapse (Bell 1989). In other words, transience in a system can be seen as a non-stationary behavior when parameters vary with time making them highly dependent on initial conditions and assessable in the context of basins state of attractor (Soliman 1995). What Bell proposes in his paper is not to address transience, but how to reproduce it in a model for training and educational purposes. In this paper, however, the purpose is not to present a model for replication of transience, but to present, how this can be assessed.

Transience within a system or situation is defined here as any change – internally or externally generated change – to the existing reducible/irreducible parameters within a predetermined range of time that alters in any way the stationary conditions of a situation. In this system/situation values are no longer known within time, but known as probability distributions at best, within the same period of time. This is not contrary to normal reasoning on situations that we as managers or practitioners do not have full control of; the most that we can say about it, is to speculate and estimate, which in our minds gives us a qualitative probability, of a possible state in a given time. Most individuals in a regular situation do not give an exact value.

Situations vary from time to time and to be able to have a list of characteristics that we as practitioners can use to identify transience becomes troublesome. Parameters that vary within situations are as unique as the same situations and depend on the reducibility of the situation. For instance, if a parameter for a situation is cubic feet of concrete needed to create a replica of the Empire State, this parameter may change over time depending on the accuracy of the calculations to know precisely the amount of concrete needed. The value of the parameter can vary and the best that we can say is that based on inflation values, the price of the materials, and labor are with 80% certainty a particular value. Again, this parameter, even though it may vary from when the project started until it finishes (making it transient within its own situation) is different than if we want to embark on the construction of a new house (it may not be generalizable).

4.3.3 Situational Understandability

Situational understandability refers to the capability of a situation to be understood. It relates to how a particular observer perceives and understands the situation as it evolves in its temporal and spatial dimensions. Bar-Yam (2005) and Biggiero (2001) mention that the complexity of a system can be based on the inherent complexity of the system or the capability of the observer to understand the situation. In this case, understanding becomes not only a part of the complex situation, but also, the element that can generate solution sets to the particular situation given that the elements of transience and reducibility be addressed by the observer.

Understanding within a situation can be defined as *the appropriate use of knowledge when appropriate* (Klahr 1974), in this case, appropriate to the situation at hand and it changes due to the evolution of it.[3] Understanding in this sense goes beyond the analysis capability of an observer. It does not only include analysis, but synthesis and synoptic capability in terms of assimilating those non-reducible ele-

[3] In this context, situational understandability is not to be confused with situation awareness (Endsley et al. 2003). Situation awareness was designed to address systemic problems in terms of reducible problems within a stable-state (Endsley et al. 2003). In addition, situation awareness is monotonic in nature, meaning that all the information that is collected of the system keeps on aggregating. Situational understandability as it is conceptualized is oriented towards dealing with complex situations addressing their transience and irreducible conditions in a nonmonotonic fashion, meaning that all info that is collected may increase or decrease the understanding of the situation.

ments within a situation. That statement, *appropriate use of knowledge when appropriate* suggests a heuristic that is unique to each individual, which relies on the perception of the observer and is based on those observations that select the right type of knowledge to take an action based on incomplete information.

According to Canon-Bowers and Salas (2000) decision making in dynamic settings is more perceptual than conceptual, in other words, the action taken based on understanding goes well beyond the analysis of the situation and more on the perception of those reducible and irreducible conditions of the situation and the transience of the situation. This perception allows the observer to select the appropriate knowledge or appropriate belief that may change once new information is made available. This reasoning mode is called nonmonotonic. Non-monotonic reasoning is based on incomplete information and changes in understanding as the situation evolves. Deductive [or inductive] reasoning cannot help in these situations (Bochman 2001).

An observer may be better suited to deal with certain complex situations given their individual capability of dealing with transient and reducibility issues. For instance, highly analytical observers may be able to deal less effectively with issues that are non-reducible, while highly reducible domains may be more complicated for abstract thinkers.

Figure 4.3a, b show the domain complexity of complex situations in terms of reducibility and transience.[4] It basically states that the more transient (T) and irreducible (Ir) a domain is, the higher the uncertainty factor ; conversely, the more reducible (R) and intransient (In) the less uncertain. This suggests that complexity may not be solely characterized by the number of elements and relations, but how reducible and transient a domain is. We may have a case of a large number of elements and relations, but relatively simple and vice versa, few elements and relations, but highly complex. It can also be observed that a transient-reducible domain and intransient-irreducible have a medium-high level of uncertainty.

By this being the gray area the most that can be said is that the one with the highest level of uncertainty will truly depend of how an individual deals with temporal-spatial dimensions. Some individuals will deal better with transient conditions, others with irreducible conditions while either reducibility or intransience conditions are present. However, it is suggested that as individuals we may be able to deal better with stable conditions. With irreducible-intransient situations we could use aggregating techniques that could overcome the analysis and establish patterns, whereas reducible-transient conditions give place to chaotic (or almost) behavior.

4.4 Non-monotonic Synopsis

Non-monotonic synopsis (NS) is basically a framework that allows for the assessment of a complex situation based on the elements of *reducibility* and *transience.* This is done by first *identifying* these elements, by *classifying* them, and finally

[4] This representation assumes that the behavior is fundamentally better able to deal with intransient, reducible conditions.

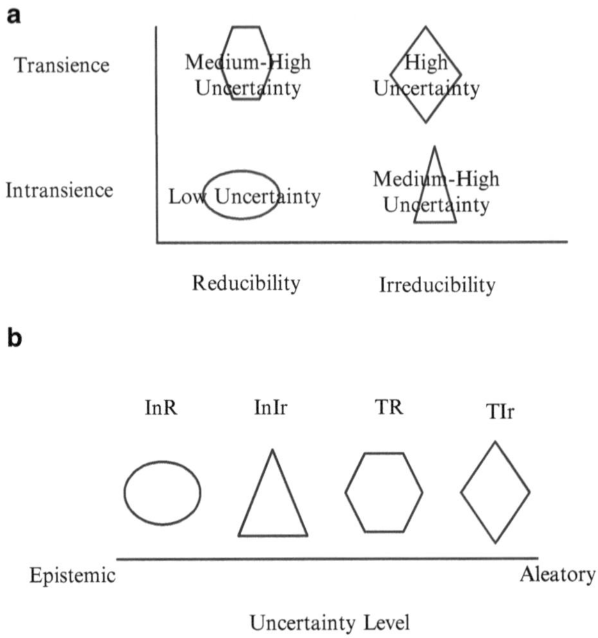

Fig. 4.3 (**a**) Domain uncertainty matrix and (**b**) Uncertainty level

assessing them. The *identification* of these elements may sound trivial, however, most professionals recognize them, but do nothing about it. We always talk about contextual issues, but little is done to approach them. We tend to identify those reducible elements first and assume that they are stable which generates problems in planning. Some of these assumptions are true to technical problems, but do not apply to socio-technical problems. The identification of some parameters that may have an effect on the situation must be brought forward regardless if at the moment there is no way of assessing them.

After what's basically a brainstorming session (with the acknowledgement that not everything can be directly measured and some instability) we proceed to *classify* the identified parameters. This classification would give four vectors: one with reducible parameters, the second with how stable those reducible parameters are, the third with irreducible parameters, and the forth with how stable those irreducible parameters are. As mentioned previously, most analysis is done with a vector of reducible parameters. What is suggested here is to consider parameters that are irreducible and that even reducible parameters may be affected by unstable conditions. Time as a dimension and for analysis purposes can be comprised and stability assumed, but for short periods of time in most cases. If stability cannot be guaranteed, in must be included as a condition under which the observer would adhere to. The non-stability condition then creates the transience vector for reducible and irreducible parameters. This transience reflects the non-monotonic nature of the framework given that it must adjust when the situation changes or becomes stable. The irreducibility of the approach reflects the synoptic need in order to study the situation at a high level without getting to low levels of analysis and losing the behavior of the whole.

Finally, the *assessment* of these parameters sheds some light in the effect that they have on a complex situation by showing how an individual understands them and somehow quantify them. Either an individual assess via analysis or judgment. There is a sense of justifiability using analysis, sense that is not found when using judgment. How it is judged is important in order to establish some transferability in the judgment structure or at least start a consensus into how a judgment was reached or how to get there. In terms of reducible conditions there are many accepted analysis practices and engineering is mostly based on them but when it comes to judgment we enter into areas that engineers are not comfortable with. To bridge the gap and to quantify some qualitative characteristics fuzzy logic as a method is suggested. While others methods are being explored, fuzzy logic is suggested for the assessment given the ease of quantifying qualitative assessments.

4.4.1 The NS Index and Fuzzy Logic: Towards a Fuzzy Expert System

The *NS index* gives information not only on how an individual categorizes the considered parameters and how this categorization compares to other individuals' categorization, but also an aggregate that establishes the perceived situational understandability of the complex situation. For instance, if for individual *a* there is predominance of reducible parameters compared to individual *b* that sees more relevant the effect of irreducible parameters it establishes the predispositions of both individuals while giving them an assessment of their perception based on the selected parameters using fuzzy logic.

Fuzzy or multi-valued logic was introduced in the 1930s by Jan Lukasiewicz and championed by Lofti Zadeh in 1965 (Negnevitsky 2002). Fuzzy logic is multi-valued and deals with degrees of membership and degrees of truth employing a spectrum of values accepting categories that can be partly true and partly false (Negnevitsky 2002). Classification of reducible/irreducible and transient/intransient parameters is undertaken by the observer, but the aggregation in order to obtain an index requires fuzzy inference. A fuzzy expert system could also be developed that contains fuzzy characteristics of most dealt with parameters within an organization.

According to (Negnevitsky 2002) there are several methods for fuzzy inference that vary in computing time, but their steps can be summarized in few steps: fuzzification, given either reducible values a fuzzy value or establishing fuzzy values for irreducible parameters and their corresponding transience level; rule evaluation, which takes fuzzy inputs and apply them to the antecedents of the fuzzy rules; aggregation of the rule outputs, which is to unify all the outputs into a single fuzzy set and finally defuzzification, which is a crisp number generated from the fuzzy system. Combining this heuristic with the identification, selection and assessment we have:

- Identification of parameters, $\mathbf{\Omega(k)} \in \mathbf{R}^n$ is a time-varying vector of parameters. It contains all the different parameters that an individual can use, at time *k*, to identity/represent a situation. All the parameters are available to every individual. However, a particular individual may use only a subset (reducible, irreducible or both) of the parameters to asses the situation. ***k*** presents the evolution in time of the situation and reflects the situational transience. In this case we could also use another vector that reflects the transience of the selected parameters
- Possible parameters
  - Internal: Financial, culture, stakeholders, requirements, internal policies, operations, timelines, personnel, resources,
  - Possible transience on internal parameters: change in requirements, change in timelines, change on policies
  - External: Technology, laws, politics, economy, demographics, social values, imposed timelines
  - Possible transience on external parameters: change in the economy, change in the political environment,
- Classification of parameters, $\mathbf{\Omega(k)}_r \in \mathbf{R}^n$ is a time-varying vector of reducible parameters while $\mathbf{\Omega(k)}_i \in \mathbf{R}^n$ is a time varying vector of irreducible parameters. Comparing the dimension of the vectors $\mathbf{\Omega(k)}_r$ and $\mathbf{\Omega(k)}_i$ and considering their weight given by the observer, we can establish, as previously mentioned, the predisposition towards reducible of irreducible constructs and the individual suitability to deal with certain situational conditions. These parameters can also be compared to an individual profile to see his/her predisposition to the selection of certain parameters.
- Fuzzification of $\mathbf{\Omega(k)}_r$ and $\mathbf{\Omega(k)}_i$
- Establish fuzzy rules $\mathbf{\Omega(k)}_r$ and $\mathbf{\Omega(k)}_i$ and how they contribute to understanding
- Aggregation of $\mathbf{\Omega(k)}_r$ and $\mathbf{\Omega(k)}_i$
- Defuzzification to arrive to NS index

It is important to re-emphasize the existence of different perspectives when assessing complex situations. When different outputs are compared, participants' initial conditions will be different, generating different parameter selection. If they were the same or planned the same, the idea is to compare parameters and focus on the elements where observers diverge while establishing a rule base towards building a fuzzy expert system. This establishes second level learning among participants by presenting their value premises within the current situation while establishing standardization in the parameters that more likely affect the organization, the type of parameters and the possible effect of the parameter.

4.5 Future Work and Conclusions

In this paper, a theoretical framework for assessing wicked problems within a context of complex situations is presented. By establishing the reducibility/irreducibility and transience of most relevant parameters and their framing, a synoptic understanding of the problem can be developed.

The importance of understanding the nature of the problem is highlighted by the incompleteness of knowledge and understanding of the wicked problem. The incompleteness of knowledge and understanding can be accepted as an attribute of wicked problems. Conversely, this incompleteness is foundational in the formation of the wicked problem within the complex situation.

The self-referential nature that is encountered implies that the ten characteristics proposed by Rittel and Weber (1973) play a dual role. They are at once reasons given for why the problem is wicked, while simultaneously they can be viewed as attributes that emerge in when problems become wicked. Characteristics such as the no stopping rule are both a cause and a result within the complex situation.

The importance of the Non-monotonic Synopsis to wicked problems lies in its ability to provide a synoptic formulation of the problem that increases the understanding of the nature of "wickedness" that must be dealt with.

The framework requires further validation to establish its scientific value and most importantly, its usability for academics and practitioners. Given the nature of the framework, a pragmatic validation is being considered instead of an empirical one due to the qualitative nature of the framework. Also, other approaches besides fuzzy logic are being pursued such as hybrid system methodology and the use of colored Petri nets could be of great use to reflect the effect of reducible and irreducible components.

4.6 Some Thoughts

As this section is an addition to the original paper, it briefly adds new thoughts to the discussion on complex situations. First, it challenges the need of the concept of complex situations as it does not necessarily generate a new explanation. Instead, the concept of problem situation is used as it is a good candidate for the general case for which complex situation is one particular case. Second, the concept of understanding is introduced as a way to assess problem situations, and finally, concluding remarks are presented.

4.6.1 Again, What Is Complex Situations?

A complex situation, as defined, has one limitation: the definition does not say what a complex situation is. By calling complex situations as "those from which wicked problems are derived that will often include both reducible and irreducible components", the authors are saying that complex situations – in one way or another – are a source of wicked problems. However, this is a description of what complex situations do, not what they are. Further, the account of wicked problems being derived from complex situations is more a description than a definition.

The model provided in Fig. 4.1 provides a good approximation for a definition. Complex situations, in this case, could be defined as the result of the combination of domain, solution, and observer. This result should consider the variety of possible solutions, different perceptions of the domain by different observers and even different perceptions by the same observer whose different possible combinations most definitely would qualify as complex, especially if we consider each in terms of degree. In other words, what makes a *situation* complex is the variety and likely the impossibility of assessing the output of such combination of factors. But, is complex situations a new concept or construct? Is it needed? Does it explain something that was not explained before?

Considering Rittel and Webber's account, a complex situation can be seen as a wicked problem. Wicked problems, for instance, do not have a unique solution (many solutions are possible), perception of them depends on the observer, and the domain is wicked by association. In other words, complex situations components can be found on Rittel and Webber's wicked problems description. A more general concept is that of problem situations (Vennix 1996). Let's depart from Arthur's account of a situation (Arthur 2000). He presents that:

- When problems are not well-defined, agents face not a problem, but a situation
- To deal with situation it must first be framed as a problem
- Framing the problem is perhaps the most important part of the solution

This account of a situation is still consistent with wicked problem as framing the problem *is the* problem. A problem situation has one more subtlety besides that of framing the problem: there are people for whom there is no problem. In other words, a problem situation is a wicked problem when people agree there is a situation which is difficult to formulate as a problem *and* when there are people that don't see the problem at all. Being this the case, the problem is of not only defining the problem, but identifying it first. At this stage a solution is not even necessary as a problem has not even been identified; therefore a complex situation cannot be formed.

4.6.2 Understanding and Problem Situations

Padilla (2010) suggests that for a problem situation to take the form of a problem it needs to be understood as a problem first and its understanding does not require a solution. In other words, do I understand that there is a problem? If the answer is no, then there is nothing to understand. Padilla proposed that understanding within problem situations is a function of an individual's knowledge, worldview and perception of a problem situation. He proposed that when the three are matched understanding occurs by assigning a truth value to the problem situation. In other words by answering the question *is s*[5] *a problem*, the individual understood that the problem situation is a problem when *s* is assigned a truth value of *true*. If the truth value assigned is *false*, then that problem situation is no longer a problem. Padilla noted that how the problem is formulated and the correctness of the formulation cannot be evaluated as any formulation is possibly correct within a problem situation. All that

[5] *s* is a statement or assertion that can be assigned the Boolean value of true or false.

a person can say is that he or she understood or did not.[6] In well-defined problems, understanding can be assessed by matching an explanation with a previously known and accepted solution, but this is not the case in the general case within problem situations as there is not a unique and accepted explanation of a problem.

One important note from Padilla's account is the use of understanding instead of knowledge to deal with problem situations. He proposes that the awareness that a worldview is at play as an axiomatic structure under which a problem situation is evaluated and under which knowledge is applied is key on realizing how understanding takes place. Further, he shows (using data from a simulation) that if the appropriate worldview is used even little knowledge can provide greater understanding than lots of knowledge with an inappropriate worldview. In other words, the traditional account that the more knowledge the better for understanding does not apply under problem situations. It may be better for problems under bounded conditions under which they can be well defined.

4.6.3 Understanding and Complex Situations

As a problem situation is the general case of a complex situation, understanding should provide an account on how to deal with them.

The NS index provides an initial step into categorizing the problem.[7] This is important as after we understand that there is a problem, now a truth value needs to be assigned to that problem. For instance after understanding that undocumented immigration is a problem, a problem that can be derived from it is *walls deter undocumented immigrants from crossing the border*. We need to evaluate this assertion and assign it with a truth value of true or false. However, as it can be observed this is not a statement which an "easily" assignable truth value where everyone would agree on. What's more important is that if a truth value of true is assigned, then it becomes a possible *solution*. Despite this being an already implemented solution, problem formulation (as already suggested) is very difficult as there are more than one formulation. All of those formulations need to be evaluated through the understanding process. However, understanding is not necessary for the truth value to be assigned. It is one way to assign that truth value (perhaps the best way). Another way could be by just relying on a worldview (political worldview, for instance) without relying on knowledge, another could be relying on knowledge without accounting for cultural differences (one possible worldview), and another

[6] Padilla posits that not-understanding is a form of understanding in which a person says he/she does not understand. Understanding results from the proper matching of knowledge, worldview and problem. Not-understanding results from an improper matching. For further explanation on what 'proper' means, please refer to Padilla (2010).

[7] It is noted that Padilla uses a different problem categorization. While he acknowledges usages such as soft and hard or well-defined and ill-defined problems, he categorizes them as problems about structure and problems about behavior.

could be by an arbitrary decision. Notice on the first two, elements of understanding were present, but not all three.

One interesting extension of this reasoning is that while we needed to identify if there was a problem and what the truth value of a problem was, now we need to evaluate if the formulated solution is indeed a solution. The propose solution can be a solution to some while for others is not, leading to a solution situation. Notice that the formulation *walls deter undocumented immigrants from crossing the border* was but one formulation. For someone else that was not even a problem that needed understanding. Now that a solution is proposed, one or more individuals for which the statement was not even a problem now need to evaluate it as a solution. The problem is that now, as this solution exists, there may be many candidates for a solution in addition to those who do not see it as a solution. This is a reflection of a complex situation; now we have a solution upon which people don't agree and may not even acknowledge it as a solution.

4.7 Final Remarks

Addressing complex situations becomes very much like addressing wicked problems: anything that you try may be good or bad as an objective traceability or effectiveness measure cannot be established. Answering the question did you understand the problem, provide an insight into three important tasks: (1) Problem situation identification, problem formulation, and identification of problem type (which are, as suggested, very difficult), (2) Assessment if existing knowledge is appropriate to assign a truth value to the problem, and (3) Identification and awareness of the worldview that drives the understanding process. Despite accomplishing these tasks and after using the triple of problem, knowledge and worldview, the most an individual can say is that, he or she understood the problem. Pragmatically, this understanding cannot be assessed as good or bad. However, it provides a starting point in realizing that we need to understand something (make sure that the three are in play) and with what, *before* we provide a solution. This seems part of common sense, yet, and as mentioned, solutions can be enacted without the proper match of the three (the problem being understood) reflecting biases or just plain gambling.

It is important to note that part of what makes a problem wicked is the presence of many individuals and Padilla's account of understanding is for an individual. Yet, he provides insight into how the concept of understanding can be extended to a group. In this, he suggests that the lack of shared understanding may be a positive property. In other words, the condition of a *situation* may be positive as different understandings provide different perspectives of the same problem. The opposite would be that people fully share their understanding resulting in one individual replicated as many people are present. He suggests the need of a balance of these two perspectives as we need some common elements as to be able to communicate with one another. How much of commonality is needed? Well, that seems to be an interesting question worth answering.

References

Arthur WB (2000) Cognition: the Black Box of economics (Chapter 3). In: Colander D (ed) The complexity vision and the teaching of economics. Edward Edgar Publishing, Northampton

Bar-Yam Y (2005) About engineering complex systems: multiscale analysis and evolutionary engineering. In: Brueckner S, Di Marzo Serugendo G, Karageorgos A, Nagpal R (eds) Engineering self organising systems: methodologies and applications, ESOA 2004, LNCS 3464. Springer, Berlin, pp 16–31

Behe M (1996) Darwin's Black Box: the biochemical challenge to evolution. Free Press, New York, p 9

Bell P (1989) Stochastic visual interactive simulation models. J Oper Res Soc 40(7):615–624

Biggiero L (2001) Sources of complexity in human systems. Nonlinear Dyn Psychol Life Sci 5(1):3–19

Bochman A (2001) A logical theory of nonmonotonic inference and belief change. Springer, Berlin/Heidelberg/New York

Canon-Bowers J, Salas E (2000) Decision-making under stress. American Psychological Association, Washington, DC

Churchman C (1967) Wicked problems guest editorial. Manag Sci 14:B141–B142

Durga Rao K, Kushwaha HS, Verma AK, Srividya A (2006) Quantification of epistemic and aleatory uncertainties in level-1 probabilistic safety assessment studies. Reliab Eng Syst Saf. doi:10.1016/j.ress.2006.07.002

Endsley M, Bolte B, Jones D (2003) Designing for situation awareness. Taylor & Francis, New York

Hora S (1996) Aleatory and epistemic uncertainty in probability elicitation with an example from hazardous waste management. Reliab Eng Syst Saf 54:217–223

Klahr D (1974) Understanding understanding systems. In: Knowledge and cognition. Wiley, New York, p 295

Kovacic S, Sousa-Poza A, Keating C (2008) Complex situations: an alternative approach for viewing system of systems. In: Proceedings of the system of systems engineering (SoSE) conference, Monterey, CA

MECS (2007) Managing and Engineering in Complex Situations Conference, Norfolk, VA

Negnevitsky M (2002) Artificial intelligence: a guide to intelligent systems. Addison Wesley, New York, pp 88–89

Padilla J (2010) Towards a theory of understanding within problem situations. Dissertation, Old Dominion University, Nortfolk

Rittel H, Webber M (1973) Dilemmas in a general theory of planning. Policy Sci 4:155–169

Soliman M (1995) Dynamic bifurcation in non-stationary Systems: transitions with an unpredictable outcome. Proc Math Phys Sci 451(1492):471–485

Vennix J (1996) Group model building. Willey, New York

Waldrop M (1987) Man-made minds: the promise of artificial intelligence. Walker & Company, New York

Chapter 5
The Gestalt Imperative: A Proposition for Understanding

Samuel F. Kovacic

Abstract With so many terms available to define the same thing, it would seem nearly irresponsible to introduce yet another term [complex situation] that appears to describe a phenomenological state such as a system. However, complex situation infers both a broader meaning and imposes a differing perspective. *Complex* in this context is dependent on understanding and reality rather than observer and knowledge, and *Situation* imposes a gestalt that cannot be characterized within a singular perspective that relegates paradox to a hierarchically imposed primacy where the "squeaky wheel" gets the attention. This also infers that complex situation has no monotonic definition or each attempt at a definition by a singular perspective is by default incomplete. Therefore the perennial derivations for system: complex systems, system of systems, federation of systems, stochastic, chaotic, dynamic etc.... are no longer a sufficient descriptor for complex situation. Ergo system and its genealogy lack the constitution to define complex situations. To wit, this paper provides the premise for a situation and describes the conditions that make it complex sufficiently for syncretic study by discipline(s) in the proposed field of complex situations.

5.1 Introduction

Understanding complex situations is a journey into the history of the major influences on nature: science and philosophy, and the correlating premises that act as an anchor for study. The journey is necessary for any chance for a shift in the paradigm to complex situations. This paper has two purposes. The first: to imbibe the meaning

S.F. Kovacic (✉)
Batten College of Engineering and Technology, Department of Engineering Management and Systems Engineering, Old Dominion University, Kaufman Hall 242, 23529, Norfolk, VA, USA
e-mail: samkovacic@gmail.com

S.F. Kovacic and A. Sousa-Poza (eds.), *Managing and Engineering in Complex Situations*, Topics in Safety, Risk, Reliability and Quality 21, DOI 10.1007/978-94-007-5515-4_5,

of gestalt into the dialog of the singular discipline and their phenomenologically based descriptions. The second: to question understanding at the personal level, instead subscribing it to the domain of reality, a purposeful segue from knowledge and its epistemic implications. The practical utility for providing this theoretical jaunt into reality and perspectives is to provide a generalizable meaning to the term complex situation and remove any historically contentious definitions that may be found within individual disciplines.

5.2 The Basis for a Gestalt Imperative

Understanding complex situations is a journey into the history of the major influences on science and philosophy and the correlating assumptions that are subscribed towards describing complex situations. It is the assumptions that are drawn that help bound and define the gestalt. For each epoch this paper provides a summary interpretation and a premise that makes explicit another part of the puzzle for what is a complex situation. The paper covers diverse subject matter based on readings from scholars spanning centuries. The summation and premises from these readings are provided in the following paragraphs

In their book *Order Out of Chaos* Prigogine and Stenger (1984) provide a chronology of nature and speak to the instability of science (in western culture) with regard to the loneliness of science as described by Pascal (pg 3). "Science can only speak in terms of science and as such fails to elicit conversation in that which is not described in nature by science." The authors use Pascal as a starting point to describe bifurcation of nature. From Pascal's work the authors turn to Diderot's radical attempt to limit or supplant science with a new perspective. Diderot provides the lens for how life can be explained. His contribution to the dialog was not to refute science but to suggest that science is not sufficient for understanding life [nature] (pg 80), which is informed by something more than science. This is highlighted by an imaginary conversation with the physicist d'Alembert.; that a notion of life as depicted in the process of an egg evolving to a chicken. His point is that evolution cannot be explained solely through the organization of living matter (replacing inert matter with active matter), Diderot states that "nature must be described in such a way that man's very existence becomes understandable" (pg 83).

Premise 1: "The Tower of Babel": Understanding [and by default description] from a singular perspective is orthodox and is relegated as an attribute to the whole [nature].

By defining knowledge in its own language, science loses its discourse with nature. Science is but one of the languages necessary for understanding ergo objective knowledge is incomplete. It must be assumed in a situation that a discipline's ability to explain phenomena within the aggregate of its own boundary is myopic and integration with another discipline is problematic. Whether the failure is from the intractable nature of the individual disciplines or through the efforts of integrating misaligned perspectives generated from each discipline, the dialog will suffer from the ensuing uncertainty and the complexity that challenge decision makers. It is assumed that there is sufficient overlap between the two disciplines to overcome any integration issues, however, the gaps are generally obscured rather than addressed by the overlap. This aggregation in disciplines leaves gaps in terms of understanding, each discipline is immersed within its own lexicon and axioms that is either subsumed by the other discipline or obviates it entirely. Within these gaps lie uncertainty and with it emergent and dynamic properties that constantly change the nature for how the problem is framed.

Immanuel Kant took an antagonistic position in regards to science stating that: "science is nothing more than metaphorical statements to include ideas of life" (pg 86). The nature of his work was identifying two levels of reality as phenomenological (all that is accessible by the human mind) and noumenological (all that is not accessible but transcends from spirituality). Kant supplanted science with philosophy as true knowledge. Kant's contribution of rational thought was depicted in the Copernican Revolution which stated: "objective knowledge cannot be anything more than what it perceives". Kant reverses this by stating "the subject does not revolve around the objective but rather the object revolves around the subject ergo philosophy [transcendental] is the truer form of knowledge." (pg 87). Establishing philosophy as the dominant position in respect to science Kant was effectively able to stem the domineering momentum of science's grasp on understanding, however, this same primacy violates Diedrot's point: 'there is no one language for nature.'

Premise 2 "Kissing cousins": Primacy is established by understanding rather than knowledge.

Kant asserts a generalizable dialect in the transcendent state and effectively shifts the conversation to the gestalt but at the expense of actionable knowledge within the epistemic dualities. Again, the dyad of Philosophy and Science is lost within each language. It must be assumed that primacy be subscribed not from the knowledge gained by the individual ontology but through the understanding gained from an overarching goal [value premise] by which an evaluations for obtaining the goal can be made. A value premise unlike an attractor becomes the focal point for dialogs of all dialects to judge the merits of how they obtain the goal.

Whitehead turned the subjective experience to one of process; he accuses the math of science as being confused and wavering amidst the paradigm of three extremes: dualist, positivist, and the constructivist. Whitehead took a personal stand against science and suggests: "the conceptual field within which the problem of human experience and physical processes could be dealt with consistently and to determine the conditions under which the problem could be solved" (pg 94). Nicholas Rescher builds off of Whitehead and provides the foundational tenets for a process philosophy. Rescher (1996) states "that a person can see reality as individual elements (substantive reductionist approach) or as a collection of elements (process holistic approach)." He puts forward the laws of science are a process and we understand the laws because we are a part of them. This approach simplifies the need to digress back into a modality of science for a coherent view of nature (Rescher 2000).

Premise 3: "Process philosophy" – There is no conflict between science and philosophy, we enact it every day

A process approach avoids the intransigence of perspectives, providing a temporal path that does not rely on deterministic or stochastic modalities to affect sufficient causation to allow for understanding to occur. Instead of requiring a sequence of events to maintain a coherent perspective, or an amorphous boundary to capture unlimited random possibilities process inserts understanding through intuition as the means for causation to an end goal.

Planck's constant allowed for Newtonian physics to exist alongside corresponding quantum physics and maintain its universal generalizable nature. However subsequent to Planck's discovery was Einstein's interpretation of the effect of Planck's constant and introduced his idea of the wave and particle duality of light – that light can be both a particle and wave (pg 219). This was followed up with Heisenberg's uncertainty principle and its devastating impact on efficient cause – re-addressing science's notion on causality and certainty (Wolfe 1989) – and inserting back the notion of final cause. Bohr's idea of complementarity – that there is not a complete description of the system 'as is' independent of how it is observed. Wolf (1989) Sousa-Poza and Correa-Martinez (2005) and Brewer (2010) add their voices to the discord building a unified voice that speaks to the idea of a separation from reality.

Premise 4 "Reality (continuous) and Domain (discontinuous)": When observing there is a separation from the continuum and the disassociated construct of the domain (that which we perceive) – making statements across disciplines of the domain probabilistic.

These premises allow us to state generalizable principles to the nature of gestalt. In the pursuit of knowledge disciplines are born and evolved. The evolution of each discipline is predicated by omissions of other paradigmatic disciplines. Each paradigmatic and disparate discipline is also complementary in nature. Neils Bohr advocated that:

> evidence obtained under different experimental conditions cannot be comprehended within a single picture, but must be regarded as complementary in the sense that only the totality of the phenomena exhausts the possible information about the objects. (Neils Bohr 1949)

The mind-independent assertions of the positivist, and the mind-dependent assertions of the constructivist, are bounded by the paradigmatic imposition of the worldview or discipline of an observer. Holistic approaches are fundamentally contained within a discipline. The ability to function in a multidisciplinary manner is contingent on the degree to which shared awareness or understanding can be established from perspectives that are derived from different disciplines. In simple conditions where the axiomatic limitations of complementary perspectives are not challenged, a high degree of shared awareness is possible. The necessity of a multidisciplinary approach is, however, obviated by the correspondence of the conclusions that may be drawn by any one discipline.

In complex conditions, complementary perspectives take on a polar or dual nature, in which the establishment of a common or shared awareness becomes

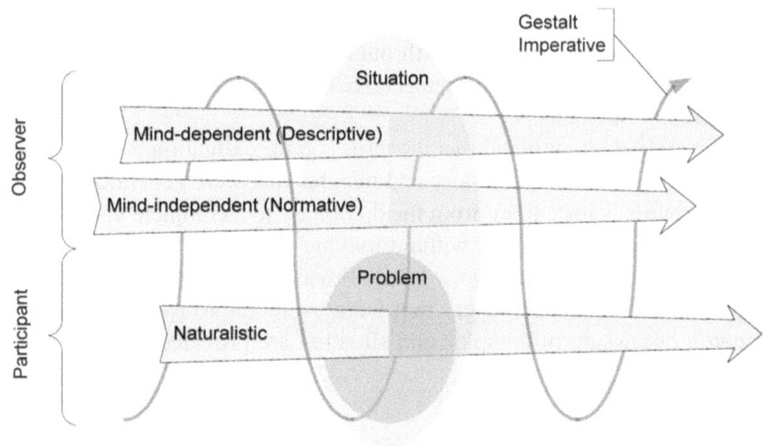

Fig. 5.1 The gestalt imperative

increasing difficult if not impossible. It is, however, in such situations, where for example under conditions of emergence a particularistic perspective finds no correspondence with a universalistic perspective, that multidisciplinary approaches are *relevant*.

A foundational concept of multidiscipline is *gestalt*. It implies that wholeness can never be recognized within the disparate disciplines that attempt to study the whole – they are too embedded within their own traditions. Instead emphasis is placed on the concept of '*meta*' (OED 2008), or universal as suggested by Plato; more comprehensive concepts that convey the idea of wholeness to its parts; either concept holds true for singular disciplines. The gestalt imperative is consequently proposed (based on the afore mentioned premises) as a canonical set of principles; its purpose is to shape a space as a multidisciplinary whole for continued study.

- Inclusivity Principle: Any discipline must accept and acknowledge complementary perspectives and dialogs. In so doing any discipline must accept its own fallacy and tendency towards orthodoxy.
- Non-Orthodoxy Principle: The Gestalt Imperative must have the mechanism embedded within it to provide the ability to let the gestalt influence the perspective without undue onus from kickback of the polar orthodoxy on the gestalt of the problem.
- Pragmatic Principle: A practical confluence where the singular or monotone decision effect becomes nearly transient is mitigated by the participatory nature of the practitioner (Fig. 5.1).

These principles provide the conduit for a gestalt imperative by providing the necessary feedback into the whole via an algedonic loop. The tension between the bi-polar perspectives is mitigated by the practical nature of the pragmatic feedback loop as well as the response from the gestalt. The significance of the gestalt imperative is that it allows understanding a problem not only phenomenologically (through observation) but noumenologically (through participation) accepting the nuance that uncertainty, which monotone perspective attempt to remove, is better accepted than extracted or obviated.

It can be argued [effectively] that the term complex situation is redundant; however, in so doing much of the history and insights that were generated to reach this state of awareness is lost. To go from the dictionary to a complete understanding of complex situation is incomplete without making the journey into the history. It is anticipated that over time the two words will become a term that, although redundant, speak to the many challenges that were overcome to recognize this nuance. Until then a discussion of complex and situation are provided as if they were not redundant words.

5.3 Complex[ity]

It is uncertain how many uses the word 'complexity' has found in the English language; as such it would be remiss not to frame how complexity is being used in this paper.

> "A key difference between current cybernetics and complexity theory is the use of different epistemology. Complexity theorists use a realist epistemology and assume that complexity exists in an observed system, or perhaps in a computer model. Cyberneticians use a constructivist epistemology and assume that the system of interest is defined by the observer." [Umpleby/WMSCI webpage]

This theme for how complexity is perceived and dealt with resonates within the research communities of many research and academic institutes, Umpleby's comments are foundational in BarYam's New England Complex Systems Institute (NECSI) where research explores both facets unilaterally. Complexity, a major division within the Santa-Fe institute has taken a multidisciplinary collaboration approach, and the University of Michigan's Center for the Study of Complex Systems encourages research in nonlinear, dynamical, and adaptive systems. Additionally, complexity is an integral thread in many centers: the National Centers of System of Systems Engineering (NCSOSE 2003) invokes complex systems in their mission statement as does the System of Systems Center of Excellence (SOSECE). Sousa-Poza (Sousa-Poza and Correa-Martinez 2005), however, opens a unique door and suggests that complexity is tied not only to the observer and how he perceives but that the observer, as a participant, is a major contributor to the complexity, insinuating both a pragmatic and fallible component to complexity.

Andres Sousa-Poza introduces complex situation in his seminal paper titled 'Pragmatic Idealism' (Sousa-Poza and Correa-Martinez 2005). The paper was the first instantiation of a budding idea that was germinating at a time with small cadres of researchers when system of systems was gaining traction in the research community. The papers' intent was to set the philosophical foundation for how to "understand and address complex situations" commuted from the idea of System of Systems. It accomplished this by establishing the relationship of what can be known (the domain) from all that is (reality) and follows up with the systemic perception of what is perceived of the domain. This was the first attempt to treat complexity and its implications on understanding rather than observation. Pragmatic Idealism (PI) provided a keystone principle used in the study of complex situations; a working definition of complexity – "complexity is proportional to the probability of having/making an erroneous knowledge claim (ρ(ε))." (Sousa-Poza and Correa-Martinez 2005; Kovacic et al. 2006; Padilla 2007; Brewer 2010) This definition recognized the edict that a systemic perception is a function of the domain, not reality, and that the approximate distance between reality and the perspective is due to the less than perfect perspective of reality through the domain, including the error in the knowledge claim as a result of this separation. This concession imbibes the idea that complexity is defined as a construct associated with the fallibility of understanding.

> Drawing on the concepts of Pragmatic Idealism presented earlier, we acknowledge that there is a reality which we cannot know. Within this reality we must define a domain on which we focus. The bounding of the domain, becomes a crucial step to reducing A(d') The distance between the domain and reality, and consequently A(d) our perception of reality. (Sousa-Poza and Correa-Martinez 2005).

Premise 5: "The one face of complexity": complexity is commuted to fallibility and the error that is generated as we try to understand a complex situation. Statements of reality are possiblistic and commuted onto the domain.

Hegel obviated the reductionist theme of science and the arrogance of speculation found in philosophy and proposes levels of complexity that correspond to the complexity in nature and to the concept of time. Hegel put forward levels that reflect the increasing complexity of nature and a concept of time that would make each level richer (pg 90). Although his philosophy never gained traction it wasn't due to the philosophical notion but rather the conditions that he builds his philosophy were made obsolete with the discovery of an alternative to classical physics – quantum. Bergson, however maintains Hegel's theme when he put forward the idea of speculative knowledge. Bergson posited that science is a whole (vice Diderot) that must be understood through rational intelligence, he states that, "rational science is incapable of understanding duration since it reduces time to a sequence of instantaneous states linked by a deterministic law" (pg 92). Bergson chose to avoid the conflict between philosophy and science in favor of something new, a philosophy that chose to address the problem of time and complexity in favor of intuition.

Premise 6: Conceptual Bounding – complexity is bounded with the conceptual boundary– nature has its own philosophy elevating science and philosophy to a new dialect of intuition, however, nature is generalized to the coexistence of science and philosophy rather than unto itself.

The necessity of simultaneously maintaining multiple disciplines can only be argued from the position of their indispensability to address a [complex] problem. For this, we must establish the limitation of an orthodox, [single discipline] position.

Within a discipline, the bounding of a problem is dictated by the principles and axioms that underlie the discipline itself. This act of bounding, however, not only influences the perspective(s) that can be supported, but the very manner that the problem and reality are perceived. The discipline in this sense will become "the hammer that makes everything look like a nail". The highest degree of comprehensibility will be marked by the nature of the bounding. Optimizing within this bound will maximize the understanding that is generated by a perspective, but will, based on the theory of complementarity (Rosenfeld 1961), become increasingly polarized and inaccessible by other perspectives. In the problems where a satisfactory solution is identified within the comprehensibility that a perspective can provide, an orthodox position is warranted. If such a solution is however not possible within the constraints imposed by the comprehensibility of a perspective, adopting an alternate perspective, as is the case in reframing, might be possible. For truly complex problems any perspective will provide a local perspective, but will be unable to generate a sufficiently global construct to generate a suitable basis for further action. Thus, a paradoxical condition is set where multidisciplinary approaches can be enacted where they are not required (simple problems), and become impossible to adopt where they are necessary (complex problems).

5.4 Situations

The question of dealing with knowledge in two dimensions [science and philosophy] generates unique issues. The duality suggested by Einstein and the treatment of the observer by Heisenberg's and Bohr become significant in the construct of complex situations. This thread necessitated a construct of knowledge that has a corresponding component to 'not-'. Padilla elaborates on the complex situations model described by Kovacic et al. that in effect:

> reflects the entity of the situation in a temporal and spatial scale, but also associates the solution form to the capability of understanding through the observer's personal profile. (Padilla et al. 2007).

This suggests that knowledge as an irreducible and transient condition is imposed by predispositions of the observer in complex situations. It is the knowledge from the domain combined with the non-knowledge of the observer towards a value premise that creates the condition for understanding.

Premise 7: "Duality of Understanding". Knowledge and not-knowledge form the basis of understanding.

Complex Situations Paradigm (CSP) is an internally consistent philosophical foundation for complex situations (Brewer 2010). Brewer threads Pragmatic Idealism, proposed by Sousa-Poza et al., inextricably into the foundations of the paradigm. CSP provides the epistemic, ontological axiological and methodological principles necessary to be labeled a [emerging] paradigm Important within his work is the definition of Cognitive Representation of Reality (CRR). CRR is: " are representation of reality in the cognitive domain that acknowledges knowing refers to something known and fundamental in comprehensibility of reality." The value premise principle forms the basis of action based on comprehension. Brewer also provides a definition of a situation, specifically: "that a construct to frame discussions of complexity relative to reality, or a portion thereof; a self-aware individual (one or more), and the individual's CRR"

Premise 8: "Comprehensibility – Access to the Big Picture". The CRR is comprehensible to the phenomena and its corollary noumena can be induced thus justifying action

Ilya Prigogine puts forth a compelling case for not stepping of the narrow path of chaos into science and philosophy. Whether or not this path set forth by Ilya is aligned with ours remains to be seen, certainly the 'cadre of researchers' pursuits for solving for complexity mirrors Prigogine's philosophically grounded, practically informed approach in the search for understanding. It remains to be seen what doors Complex Situations Paradigm will open but, it certainly is comforting to know that the narrow paths may be many, each one with a equi-probable chance of getting to the end.

5.5 Implications to Complex Situations

The practical implications of these assumptions for a complex situation is the separation from the pursuit of knowledge (research) and practice (action) will increase exponentially the degree of error in the knowledge claim. The pursuit of science that is consistent and generalizable is contrary to practical behavior of a group and their random patterns or the specific action of an individual and its impact on the whole.

This condition generates issues such as.

- Non-monotonic – a condition where the knowledge is tenuous due to incompleteness, so much so that understanding is low and the commitment to 'doing' is limited. In a non-monotonic situation understanding does not increase with the increase in knowledge.
- Indefeasible – in a condition such as non-monotonic where 'doing' is sufficiently questionable that the actor will reserves the right to commit based on the chance that more information will become available.
- Multiplicity – there exists multiple CRR's of which some CRR's will have a shared domain of awareness and others may not, rendering a dialog impossible.
- Emergence – the behavior of the whole [emerges] is not the sum of its parts, with emergence is non-ergodicity – in the absence of states transitions between states cannot be made explicit.
- Uncertainty – with fallibility there is uncertainty, rendering useless absolute statements. With uncertainty comes the 'equi's' (equal likelihood's)
  - Equiprobable – equal valid alternative ways of attaining the same objectives (divergence).

  - Equifinality – from a given initial state, obtain different and mutually exclusive objectives (convergence).
- Irreversibility – the result of approximation forever breaks the path back from effect to the cause, yet with the laws of nature we know with certitude that once initial conditions are set forth everything is determined.

These are the implications of complexity as they are researched within a complex situation. The foundational concept of researching what is not known to deal with what is known is unique. Within the non-monotonic nature of complex situations the ideas of techniques, methods, or process become problematic because there is no way to judge them towards the final outcome, at best we can state that they are models for which pragmatic understanding can be affected.

It is the predictive nature of monotone process that limit its use in complex situations the implications would suggest that using probabilistic modeling for contribution to transient and intransient information would greatly enhance the decision makers ability to understand the problem however it is inconceivable that due to indefeasible nature of a complex situation that stochastic processes can encompass all the facets of a complex situation. Amidst the morass of all possibilities within a complex situation the perturbations of probabilities would be mind numbing.

[As such, and] In light of the assumptions put forth in this essay it would be premature to ascribe the term system to complex situations; it would be a misnomer. As a model for generating and manipulating knowledge (random, linear, and non-linear) within a state has made its mark, but fall short of addressing all the implications of a complex situation.

Despite the coherence and conviction to justify the promise that Deterministic Chaos brings to understanding nature, Ilya still felt compelled to insert the platitude to the history, destination, and most importantly the determination of nature in regards to the dominant discourse of science and philosophy. Ilya's angst over the dominant discourse serves as a warning to complex situations that the merit of complex situation is harbored in our readers not our own convictions.

5.6 Conclusion

A complex situation is rational process that is fallible in construct, embodies all possibilities within a conceptually bounded space for the purpose of understanding within time independent action. Prigogine predicts the end of certainty in the advent of a deterministic chaos movement, a movement that would propel stochastic methods of probability to the forefront for explaining nature. Prigogine admits to the 'narrow path' he has taken between science and philosophy, and the path may well lead us to ultimate truths, however there is still the nagging issue of transformation and change. Nassim Taleb's *Black Swan* (Taleb 2007) – reminds us that time is an issue that has not been addressed; and time, as Prigogine suggests, may be the influence and/or dogma of the great Greek writers. However, Pragmatic Idealism give us the framework to perceive sans time and it is through our perception of the domain that we form solutions; complexity, determinism, and chance exist within the realm of the domain as an abstraction that is constantly being informed by reality. As such complex situation does not raise the banner for any singular language over the other, but provides the means for potentially a new paradigm over time.

References

Bohrs N (1949) Discussions with Albert Einstein on epistemological problems in atomic physics. In Albert Einstein: philosopher-scientist (1949). Cambridge University Press, Cambridge

Brewer V (2008) Anatomy of systems: a debate on complexity candidacy exam. Old Dominion University, Norfolk

Brewer V (2010) A decision making construct for complex situations. Dissertation, Old Dominion University

Kovacic S, Sousa-Poza A, Keating C (2006) Complex situations: an alternative approach for viewing a system of systems. In: Systems of systems engineering/2006 IEEE international conference, 26–28 April, Los Angeles Ca

National Centers for System of Systems Engineering (NCSOSE) (2003) Mission statement. Old Dominion University, Norfolk. http://www.ncsose.org/

Padilla J, Sousa-Poza A, Tejada A, Kovacic S (2007) Towards a diagnostic framework for understanding complex situations. In: 7th international conference on complex systems, Quincy, MA, 2007

Padilla J, Sousa Poza A, Tejada A, Kovacic S (2007) Towards a diagnostic framework for understanding complex situations. In: Proceedings of the 7th international conference on complex systems, 7th international conference on complex situations, Quincy, MA

Prigogine I, Stengers I (1984) Order out of chaos, Man's new dialogue with nature, Bantam new age books. Includes works from: Pascal B 1623–1662, Diderot (1713–1784), Kant I 1724–1804, Hegel G 1770–1831, Bergson H-L 1859–1941, WhiteHead A 1861–1947

Rescher N (1996). Instructive journey: an autobiographical essay. University Press of America, Lanham

Rescher N (2000) Process philosophy, a survey of basics issues. University of Pittsburg Press, Pittsburgh

Rosenfeld L (1961) Foundations of quantum theory and complementarity. Nature 190:384–388

Sousa-Poza A, Correa-Martinez (2005) Pragmatic idealism as the basis for understanding complex domains: the trinity and SOSE. In: 2005 IEEE international conference on systems, man and cybernetics, vol 3, Anchorage, AK, 10–12 Oct 2005, pp 2744–2750

Taleb N (2007) The black swan: the impact of the highly improbable. Random House, New York

Wolfe F (1989) Taking the quantum leap: the new physics for the non-scientist. Harper & Row Publisher, New York

Part II
Description of Wicked Problems

Chapter 6
Sustainable Development as a Wicked Problem

Jonathan Pryshlakivsky and Cory Searcy

"...what progress denigrates and if necessary punishes are any serious attempts to focus on an integrated view of human actions."

John Ralston Saul (1995)

Abstract This concept paper introduces the notion of sustainable development (SD) as a wicked problem. The characteristics of a wicked problem are reviewed and briefly discussed as they apply to SD. Sustainability is marked by a high degree of stakeholder subjectivity. Its long history can invoke multiple concepts such as resource scarcity, conservationism, environmentalism, or a business model. Because of the different levels of complexity and extensive network of stakeholders, sustainable development makes for apt designation as a wicked problem. It is also proposed that wicked problems can be reduced to two basic elements: kinetic and subjective. Because of the complexity involved in tackling SD as a wicked problem, the usage of Systems of Systems (SoS) platforms are briefly described and deliberated.

Keywords Sustainable development • Sustainability • Wicked problems • Complexity

J. Pryshlakivsky
Environmental Applied Science and Management Program,
Ryerson University, Toronto, ON M5B 2K3, Canada

C. Searcy (✉)
Department of Mechanical and Industrial Engineering,
Ryerson University, Toronto, ON M5B 2K3, Canada
e-mail: cory.searcy@ryerson.ca

S.F. Kovacic and A. Sousa-Poza (eds.), *Managing and Engineering in Complex Situations*, Topics in Safety, Risk, Reliability and Quality 21, DOI 10.1007/978-94-007-5515-4_6,

6.1 Introduction

The terms "sustainable development" and "sustainability" came into widespread use following the 1987 publication of *Our Common Future* by the World Commission on Environment and Development (WCED). Although myriad definitions have been offered since then, that report provided the most cited definition of sustainable development (SD): "development that meets the needs of current generations without compromising the ability of future generations to meet their needs and aspirations" (WCED 1987). However, while that definition provided a framework for thinking and discussion about SD, efforts at the national, regional, and organizational levels over the last two decades have demonstrated that implementation is hardly facile. One of the reasons SD is so difficult to achieve in practice is that it is a "wicked problem". Like all wicked problems, SD issues are often characterized by a lack of clarity, uncertainty, ambiguity, high risk, and limited understanding. Among other challenges, these characteristics make establishing appropriate analytical boundaries problematic. Moreover, the fact that SD necessitates addressing environmental, economic, and social issues simultaneously further complicates the matter. In this light, the purpose of this paper is to introduce the notion of SD as a wicked problem. Particular emphasis is placed on considering issues associated with SD at the regional level. The organization of this paper is divided into three general sections. The first deals with the history and contextual nature of sustainable development. The next section elaborates on the origins of Rittel & Webber's thinking with regard to wicked problems and the characteristics which mark wicked problems. The third section explains how sustainable development displays many of the classical traits of a wicked problem. In this last section there is also a brief discussion of regional sustainable development and the possible merits of System of Systems (SoS) approaches.

6.2 Sustainability: A Diffuse Paradigm

Sustainability is not new, nor is it even modern. Naess (1977) contends that sustainability can be traced back to Spinoza's *Ethics* (1677), wherein Spinoza claimed that man is merely an element of the universe by the unity of substance (i.e., matter), rather than the world relegated to a subset of man's reality to be manipulated to suit his own varied ends. Yet, Spinoza may not have been alone in recognizing this truth. British historical geographer Richard Grove (1990) noted that the emergent mercantilism of the fifteenth century caused enough damage so that by the middle of the seventeenth century:

> …a coherent and relatively organised awareness of the ecological impact of the demands of emergent capitalism and colonial rule started to emerge and to grow in a fully-fledged theory about the limitability of the natural resources of the earth and the need for conservation. (pp. 19–20)

During the nineteenth century, both Thomas Malthus and J.S. Mill deliberated the problems of uncurbed and exponential population growth in lieu of a resource base

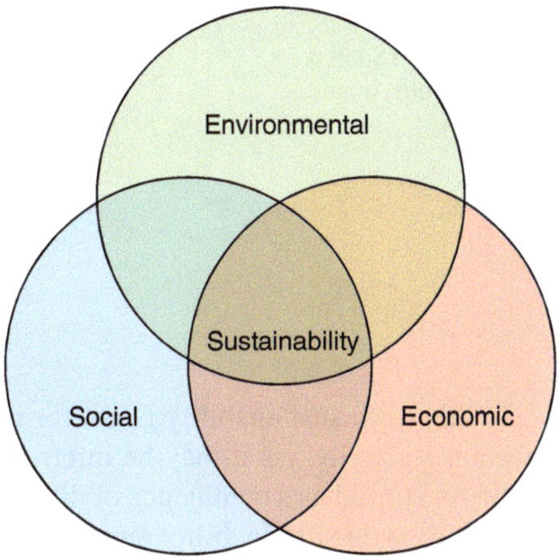

Fig. 6.1 Classical sustainability (Source: Adapted from Adams 2006)

which is linear in its ability to expand production. At the same time, noted economists such as Karl Marx and David Ricardo considered resource scarcity and its implications (Perelman 1987; Mosselmans 2004). The potential economic limitations imposed by resource scarcity continued well into the next century and beyond (e.g., the important contributions of *Rent under the assumption of exhaustibility* (Gray 1914), and *The economics of exhaustible resources* (Hotelling 1931)).

As an ecological principle Basiago (1995) states sustainability emerged from mainstream sources such as *Blueprint for survival* (Goldsmith and Allen 1972) and *Limits to growth* (Meadows et al. 1972). However, it would be remiss to suggest that a nascent environmental movement had not already emerged since the publication of Rachael Carson's highly influential *Silent spring* (1962). By this time, there had been already a 100 year history of conservationism within the United States (O'Riordan and Viosey 1998). For example, the first national park established in the United States was Yellowstone National Park in 1872.

What is most current, perhaps even fashionable, about sustainability is its emergence from near obscurity into the forefront of many aspects of activities and planning, be they economic, social or otherwise. In etymological terms, the modern concept of sustainability is intrinsically tied to concepts of ecology as a limiting factor to economic development; hence, why the phrase 'sustainable development' is the most common mot occurring in literature. Although not explicitly referenced as such, the concept of sustainable development was cited in principles 3 and 5 of the Stockholm Conference of 1972 (Frey and Yaneske 2007). The phrase "sustainable development" *proper* originates from the Brundtland Commission Report of 1987. As noted in the Introduction, the typical definition of sustainable development used from the Report is, "development that meets the needs of the present without compromising the ability of future generations to meet their own needs" (WCED 1987, p. 8).

Two emergent views of sustainability are now in the main. Classical models of sustainability typically invoke the Venn diagram paradigm as seen in Fig. 6.1 (Agyeman and Evans 2003; Allan et al. 2008; Saha and Paterson 2008).

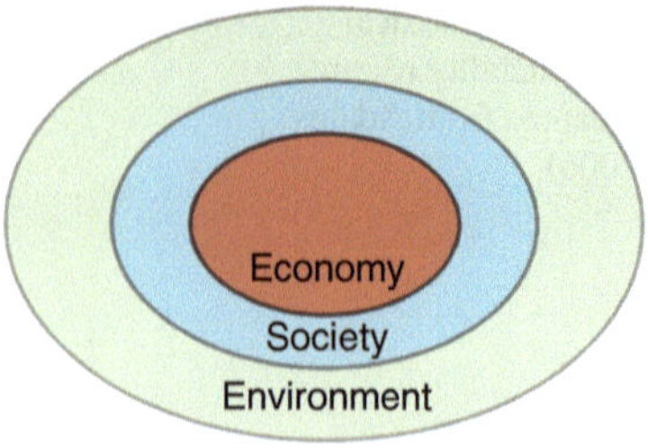

Fig. 6.2 Strong sustainability (Source: Adapted from Adams 2006)

In this case, sustainability is not for its own sake, but assists in the continuity of human activity. As it may be inferred from the diagram, the issue of sustainability is viewed as a confluence of different interests: societal, environmental and economic, with sustainability merely a subset of these elements.

The dissatisfactory state of affairs with regard to models of sustainable development, as well as the general lack of progress in this area with its implications of green washing (Parr 2009), has led to the differentiation between strong and weak sustainability (Ott 2003; Neumayer 2010). The former has been linked to the concept of deep ecology. The model of strong sustainability invoked here is of the Venn diagram paradigm, but with a different orientation (Fig. 6.2).

Here, the structural model is rearranged so that sustainability is not merely a facet of human activity but, in order to qualify as sustainable, all human activity must be constrained by limitations set by the environmental carrying capacity. Similarly, this implies the maintenance of constant levels of natural capital, which is made up of non-renewable resources, systems which maintain renewable resources and natural systems which absorb anthropogenic emissions and pollutants (Jabareen 2008).

Defenders of the classic model, such as Holling (2001), suggest that sustainable development is not an oxymoron, but is the amalgamation of two separate yet complementary concepts, what Holling terms a "logical partnership" (p. 399), perhaps to suggest that sustainable development is a zero-sum game. Holling defines sustainability in terms "to create, test and maintain adaptive capability"; that is, sustainability is limited to the capacity of the environment, a rendering somewhat similar to deep ecology. However, Holling in turn defines development as "the process of creating, testing and maintaining opportunity" (p. 390). It might be useful, though, to point out that nowhere in Holling (2001) is there mention of externalities arising from the pursuit of opportunities. Other optimistic examples of what constitutes sustainable development abound. For example, Hadorn et al. (2006) supply the following:

> The model of sustainable development is about change in social institutions in the direction of more equity within and between generations in opportunities to meet needs. It takes account of global change and its harmful effects on people, due to the complex interrelations among ecological, economic and social systems. Therefore, sustainable development can be regarded as an elaboration of the old idea of the common good, extending it in accordance with the precautionary principle to the global population and possible harm in a long-term perspective. The quest for knowledge in society is to find problem solving strategies that meet the needs of all affected parties. (p. 122)

The determination of what constitutes a common good is problematic in itself, overlooking any attempt to reconcile it with the precautionary principle, which means something entirely different. As stated in Principle 15 of the Rio Declaration (1992), the precautionary principle is evoked under circumstances of incomplete knowledge: "Where there are threats of serious or irreversible damage, lack of full scientific certainty shall not be used as a reason for postponing cost-effective measures to prevent environmental degradation." For example, although the scientific knowledge with regard to greenhouse gases is incomplete, such a fact cannot be used as an excuse to bar measures to limit anthropogenic emissions which contribute to the overall dilemma of greenhouse gases. However, it is still left unexplained how sustainable development can be reconciled with the precautionary principle, especially with regard to curtailing human activities which may be harmful to the environment, while maintaining the living standards and human liberties promised by sustainable development. Contrary to Holling (2001), Jabareen (2008) considers the substance of sustainable development an "ethical paradox" (p. 181). Summarizing Sachs (1993), Jabareen (2008) states that sustainable development,

> has attracted such a large following because it seems to hold out the promise of bringing about a rapprochement between ecological (sustainability) and economic (development) interests. [Sustainable development] is accordingly deemed able to cope with the ecological crisis without affecting existing economic relationships of power. Capitalism and ecology are no longer contradictory when brought together under the banner of [sustainable development]. (pp. 181–182).

For this reason, Jabareen (2008) asserts that the concept of sustainable development, complicated as it is by issues of ethics, justice and equality, is another in a long series of utopianisms (see also Frame and Brown 2008). On the other hand, Baker et al. (1997) have made what seems a useful distinction between what is termed anthropogenic and eco-centric paradigms of dealing with issues of sustainability. As explained by Espinosa et al. (2008):

> Anthropocentric [*sic*] favours an interventionist approach to socio-economic development where nature is seen as a basic resource for tackling humankind's problems. In contrast, eco-centric adopts a holistic approach based on a combination of social needs, ecological limits and quality of life, treating the natural and human social systems as co-evolving in a recurrent dance of interaction, each dependent on outputs from the other and providing inputs to it. (pp. 637–638).

This division between anthropocentric and eco-centric approaches to sustainability is indicative of the varying divisions between what is also classed as strong and weak sustainability.

6.3 Wicked Problems: Form and Substance

Horst Rittel taught design and architecture for more than 30 years at a variety of different institutes ranging from Germany to California, yet surprisingly never designed a building or practised as an architect (Rith and Dubberly 2007). Despite this impediment, Rittel's thinking to this day still courses through design literature (e.g., see Buchanan 2009). The cause of this is due largely to Rittel and Webber's *Dilemmas in a general theory of planning* (1973), which is still considered the seminal paper on wicked problems and what Rittel termed 'second generation' design. It introduced a new paradigm for thinking about complexity. Prior to this, planners typically followed a linear, rationalised method of problem solving, sometimes termed the 'waterfall model'.

Borrowing from the scientific method, the type of problem addressed by this methodology was designated a 'tame' or 'benign' problem by Rittel and Webber (1973). A succinct list of the qualities of a tame problem is listed below (Conklin 2006, pp. 9–10):

1. has a well-defined and stable problem statement,
2. has a definite stopping point, (i.e., when the solution is reached),
3. has a solution which can be objectively evaluated as right or wrong,
4. belongs to a class of similar problems which are all solved in the same similar way,
5. has solutions which can be easily tried and abandoned,
6. comes with a limited set of alternative solutions.

Thus, a tame problem entails a classical amalgamation of elements which can be drawn from positivism and, to some extent, empiricism. Under the circumstances of a tame problem, there is relatively little need to make references to experience, judgment, ideology or obscurantism. This way of viewing problems was facilitated into the eponymously named Taylorism or 'scientific management', which was introduced to the American Society of Mechanical Engineers in 1895. However, while Taylorism did enable unprecedented and massive industrial advancement during the twentieth century, it also did reach its nadir through its association with abuses of power during the Soviet and Nazi regimes (Saul 1992; Ravetz 2006).

It is not entirely possible to reject the notion that Rittel & Webber's thinking emerged as a result of a moral vacuum. This point seems probable given the subject of a guest editorial in *Management Science* penned by C.W. Churchman, a Berkeley colleague who "recently attended" a seminar given by Rittel in 1967. In the editorial, Churchman (1967) states, "operations research and management science have been largely indifferent to the morality of the profession… [and] that the profession has a moral problem" (p. B142). However, it must be said that Churchman does not mean this as a "moral reprimand", but rather implies that the denial of wicked problems (by attempts at taming) is a form of "deception" and, hence, "morally wrong" (Ibid.). Further evidence burgeons in the text of Rittel and Webber (1973) which displays its theoretical roots within this moral dilemma. On the very first page of their work there is a reference to G.B. Shaw's sweeping disdain for the professional, a sentiment which is later explained in the text: "to substitute expert professional judgment for those of contending political groups… would not necessarily make the outcomes better" (p. 169). Similarly, Rittel & Webber contemporaneously pronounce the death of Voltaire's naïve pantomime Candide, thus signalling the demise of "a deep-running current of optimism… in a guaranteed Progress [*sic*]" (p. 157) as promised by science. There is also a reference to Charles J. Hitch, chief economist at RAND and a dissenter on the value of Game Theory in strategic decision making (Kraft 1960; Poundstone 1992). In fact, this may not be surprising, given the tacit disdain for Game Theory which is an undercurrent theme of Rittel & Webber's text, with its repeated references to chess and zero-sum games. A telling piece was penned by Hitch in 1969 when he assumed the presidency of the University of California, which includes Berkley. In it, Hitch (1969) makes reference to the "complex problems" and "grave problems" which marked American society and how "the traditional intellectual response…centered in our universities…assumes that knowledge is the principal key to the solutions we so urgently seek" (pp. 4–5). Hitch was signalling a change at the University of California system, claiming "we have perhaps been more concerned with scientific than with social questions" (p. 5), a signal which seems to accord with Rittel & Webber's work.

Coyne (2005) has gone so far as to link Rittel & Webber's thinking with that of contemporary anti-capitalist, anti-authoritarian philosophers in France, namely Gilles Deleuze and Felix Guattari, authors of *Capitalism and schizophrenia* (1972). It would be haphazard, though, to suggest with any certainty to what degree the speculative radicalism of Deleuze and Guattari are paralleled in Rittel & Webber's thinking. Not surprisingly, the evidence is tenuous. Nevertheless, Coyne observes (2005, p. 6):

> Rittel and Webber joined a chorus of dissenters from [doctrinal design] rationalisation and argued persuasively, and in terms understandable to the systematisers, that the design process, and any other professional task, is only very poorly explained in terms of goal setting, constraints, rules and state space search. The skill of the professional is better expressed in the actual framing of the problem to be addressed.

Part of the reason for this opinion can be found in the closing pages of Rittel and Webber (1973) where clearly they root the designer to the political dilemmas in which they participate. This is to say Rittel & Webber envision the designer as a mere element of a total political discourse on wicked problems; as such, "planning is a component of politics" (p. 169). As Coyne (2005) expressed, wicked problems incorporate "value judgment, criticality, testing in context, uniqueness, inter-connectedness, authority of explanation, sociality and the interplay of diverse value systems" (p. 7). Others such as Rith and Dubberly (2007) have gone so far as to include politicking as an attribute in the process of design involving wicked problems. From their egalitarian evocation of stakeholder participation and mistrust of expert knowledge, such characterisations have been tacitly expressed in the interest of democracy.

From a more practical perspective, wicked problems emerged as a subset of twentieth century fields such as post-Newtonian, post-normative and proscriptive science. Normal science stems from Thomas Kuhn's concepts of paradigm shifts and a positivistic science characterised by "universal, objective and context-free knowledge" (Haag and Kaupenjohann 2001, in Frame 2008, p. 1114). On the other hand, post-normal science entails, "assumptions of unpredictability, incomplete control and a plurality of legitimate perspectives" (Funtowicz and Ravetz 1993 as cited in Frame and Brown 2008, p. 227). As such, post-normal science is

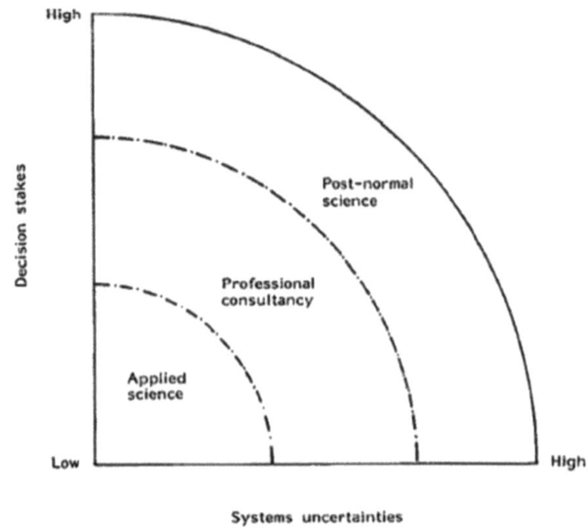

Fig. 6.3 Post-normal science (Source: Funtowicz and Ravetz 1993 Reprinted from Futures, Vol. 25, No. 7, Funtowicz, S.O. and Ravetz, J.R., Science for the post-normal age, Pages 739-755, Copyright (1993), with permission from Elsevier.)

"transdisciplinary, context-sensitive and committed to methodological pluralism and models of active stakeholder engagement" (p. 226). Conceptually, post-normal science has been described by Funtowicz and Ravetz (1993) in Fig. 6.3.

As such, applied science (tame problems) is marked by low levels of uncertainty and decision stakes, which consist of "all the various costs, benefits, and value commitments that are involved in the issue through the various stakeholders" (Funtowicz and Ravetz, p. 744). The intermediate level of professional consultancy requires further attention than the tame technical aspects of applied science; this is the locale of expert knowledge. It might also be noted that the dashed lines in Fig. 6.3 represent the overlapping of regions, while the solid line at the highest levels of post-normal science indicates a terminus. Therefore, post-normal science involves the highest levels of uncertainty and decision stakes. Furthermore, "post-normal science occurs when uncertainties are either of the epistemological or the ethical kind, or when decision stakes reflect conflicting purposes among stakeholders" (pp. 749–750). This feature pervades many aspects of a wicked problem. Also, "post-normal science has the paradoxical feature that in its problem-solving activity the traditional domination of 'hard facts' over 'soft values' has been inverted" (p. 750). As mentioned earlier, Rittel & Webber often contrast wicked problems with tame problems. Batie (2008) states "linear model and normal science were not challenged as long as society and policy makers felt that they were meeting social needs… [however,] it is no longer widely assumed that scientific progress leads to society progress" (p. 1180). The Rittel & Weber dynamic is egalitarian within its paradigm and it calls for increased stakeholder participation within the problem design. It may be contended that abuse of science is a key concern of Rittel & Webber's thinking.

The key properties of a wicked problem are outlined as follows in Rittel and Webber (1973); below is an elaboration on these properties, as well as some commentary about their meaning:

1. **There is no definite formulation of a wicked problem.** According to Rittel & Webber, any ability to understand a problem is dependent on the concept for solving the problem. Key to describing a wicked problem is the necessity to develop an exhaustive inventory of conceivable solutions. This is because "an image of the problem and of the solution emerges gradually among the participants, as a product of incessant judgment, subjected to critical argument" (p. 162). Thus, stakeholder participation is imperative.
2. **Wicked problems have no stopping rule**. When addressing a tame problem, the problem-solver knows when the job is done and when the problem is solved. With wicked problems, the key determinant in the cessation of work is due to factors exterior to positivist logic, such as time, money or the achievement of a state that is deemed acceptable.
3. **Solutions to wicked problems are not true-or-false, but good-or-bad.** Under similar controlled conditions, the findings of a particular scientific study should be replicable. Similarly, solutions to tame problems can be verified by external sources. On the other hand, the criteria for judging the validity of a 'solution' to a wicked problem are strongly stakeholder dependent. Different stakeholders see different 'solutions' as better or worse; that is, they involve judgments which may extend across a spectrum rather than 'either/or' binary paradigms: "[stakeholder] judgments are likely to differ widely to accord with their group or personal interests, their special value-sets, and their ideological predilections" (p. 163).
4. **There is no immediate or ultimate test of a solution to a wicked problem.** There is a kinetic and dynamic element to a wicked problem which prevents the achievement of finality or closure. Actions have consequences which can unfold in a myriad of different ways across an extended time period. Rittel & Webber contrast this with a tame problem which has limited scope and a high degree of specificity. Tame problems involve limited numbers of participants, whereas wicked problems affect cohorts, communities and populations.
5. **Every solution to a wicked problem is a "one-shot operation"; because there is no opportunity to learn by trial and error, every attempt counts significantly**. The content and depth of a wicked problem is such that its consequences are irreversible. There is a dynamic and breadth of scope characterizing wicked problems which precludes heuristics. Rittel & Webber necessitate this qualification on the basis of the length of time involved in a wicked problem.
6. **Wicked problems do not have an enumerable (or an exhaustively describable) set of potential solutions, nor is there a well-described set of permissible operations that may be incorporated into the plan.** Wicked problems are evolving and metamorphic. Because they alter over the course of time, the particular preferences for a good or bad solution might change or unforeseeably emerge. Thus, stakeholder preferences might adapt on the basis of changing needs or preferences.

7. **Every wicked problem is essentially unique.** This is not the contention that each wicked problem is entirely unique, but that there is some element of the problem which is in and of itself unique. As such, the adaptability of solutions to historically similar problems may not necessarily apply or be complete. This seemingly ahistorical approach is promoted in the sense which rejects the sweeping universality of positivism and perhaps its proscriptive nature.
8. **Every wicked problem can be considered to be a symptom of another problem.** Rittel & Webber envisage problems in terms of a vertical hierarchy whereby the complexity of a problem is in proportion to the hierarchical level. Most problems which are solved are usually within the bounds of competency, either individual or institutional. Thus, the temptation is to reduce complex problems into smaller, more easily defined and practical problems. However, the depth of a wicked problem is such that attempting to deal with it in a piecemeal manner can either be inconclusive or possibly harmful. Thus, Rittel & Webber promote solving problems at the highest possible level. As such, this requires the promotion of interdisciplinary, multi-party participation. However, it is interesting that Rittel & Webber don't extend the model laterally. Perhaps the reasoning behind this absence is that it would interfere with the notion of each wicked problem's uniqueness.
9. **The existence of a discrepancy representing a wicked problem can be explained in numerous ways. The choice of explanation determines the nature of the problem's resolution.** Like property 3, wicked problems have an element of subjectivity. The resolution of scientific problems has universality by virtue of its ability to be replicable. On the other hand, a wicked problem opens up the avenue for interpretation since "the modes of reasoning used in the argument are much richer than those permissible in the scientific discourse" (p. 166). This has the same effect of rejecting simplistic logical errors such as the syllogisms *if A then B*.
10. **The planner has no right to be wrong.** This is very similar to property 5 in that a wicked problem is not retractable and there is very limited room for recovery. Rittel & Webber cite Karl Popper's observation that the scientific method is a way of arriving at truths through the refutation of proposed hypotheses (i.e., what is termed the null hypothesis). However, there is a slight misinterpretation here on the part of Rittel & Webber. They state, "Consequently, the scientific community does not blame its members for postulating hypotheses that are later refuted…" (p. 167). Of course, this is not accurate, as might be expected from obvious anachronisms such as eugenics, which are rightly held in disdain. What Popper is describing and what is essentially happening in the scientific method is a process of negation. In scientific research (tame problems), both a null hypothesis and an alternate hypothesis are proposed. Should analysis demonstrate statistical incongruence with the null hypothesis, the scientific method rejects the null hypothesis and the alternate hypothesis is conditionally accepted. Note, there is no claim of proof here with the scientific method. Instead, results form a postulation which augments knowledge. That said, Rittel & Webber's intention here is to expound the notion of liability. By this, they mean that the consequences of error in wicked problems are not trivial.

Rittel and Webber (1973) never claimed that their system required all of the qualities listed above to qualify as a Wicked Problem. Rather, they contended that "there are at least ten distinguishing properties" (p. 160). Some have tried to trim down the original list of wicked problem properties. Conklin (in Christensen 2009) has reduced the substance of wicked problems into six key elements. Essentially, they are thematically equivalent to Rittel & Webber's properties 1, 2, 3, 5, 6 and 7. Similarly, Rayner (2006, in Frame and Brown 2008) has also reduced Rittel & Webber's properties of wicked problems to a list of six qualities. These are (p. 227):

- Symptomatic of deeper problems
- Unique problems that cannot be reversed
- Unable to offer a set of alternative solutions
- Characterised by contradictory certitudes
- Contain redistributive implications for entrenched interests
- Persistent and insoluble

With the exception of bullets three and four, there is similarity between Rayner's properties and properties 4, 5, 7 and 8 in Rittel and Webber (1973). The property "unable to offer a set of alternative solutions" is a unique property, but can be a consequence of having an ill-defined problem. At the same time, the property "characterised by contradictory certitudes" can mean opposing viewpoints arising from political dispositions. However, Ravetz (2006) sees contradictions as those which can be identified through "networks of relations" between the various elements which make up the identity of a system (p. 279). As happens with scientific management, the tendency to simplify matters consequently enforces "governance by targets." This drives a wedge between the networks of relations, causing friction from incongruent perspectives. As such, Ravetz calls Taylorism a "fantasy" (p. 280).

Although Rittel and Webber (1973) characterize wicked problems as including "nearly all public policy issues" (p. 160), this is perhaps overly optimistic and indulging generalisation. Many claims have been made and cases put forward as to what constitutes a wicked problem, some of which are moot. For example, Berardini (2007) asserts that modern warfare (that is, post-Cold War warfare), encompassing a variety of theatres of operation (Afghanistan, Iraq, etc.), is an example of a wicked problem. In many respects, Berardini may be correct. On the other hand, Batie (2008) asserts a whole assortment of issues, all constituting examples of wicked problems. These range from such diverse topics as terrorism, global climate change and trade liberalization. However, there is the difficulty that should a problem it be deemed sufficiently nebulous, it could erroneously constitute a wicked problem. Thus, it is propitious that Nie (2003) points to a worthwhile differentiation between problems which are "wicked by nature" and "wicked by design" (p. 308). This is because (pp. 308–309):

> the very nature and context of some cases and issues essentially promise political conflict—they are wicked by nature. But they are also wicked by design in that political actors, institutions and decision making processes compound them.

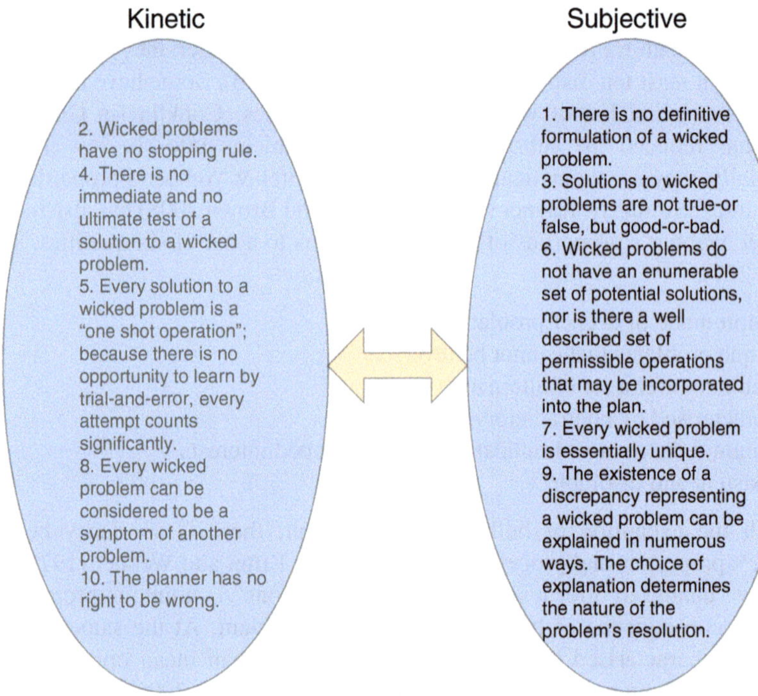

Fig. 6.4 Characteristics of wicked problems

Nie (2003) makes ample reference to the framing of a wicked problem, that these kinds of problems are "part of a larger policy story told by various political actors" (p. 320). It is precisely because of this reason that Nie claims they are intractable. Hence, Rittel & Webber's view of the designer as a "player in a political game" might result in victimization by political manoeuvring. It is for that reason that there should be some way of balancing Rittel & Webber's need for leaving open the problem definition (as in property 1) while, at the same time, providing some mechanism for preventing intentional obfuscation.

It is contentious whether wicked problems can be reduced even further than Rayner's six elements. But, in some respects there is a common current running through Rittel & Webber's original ten properties; indeed, making such an assertion seems reasonable since many of the properties appear to be significantly related to, or adjuncts of, other properties. In such a way, it may be possible to conclude that there are two central themes to wicked problems. One is that there is a kinetic dynamic to them: they are transitory in nature, metamorphic and unlikely to be resolved under conditions which are acceptable to each stakeholder. Wicked problems are also characterized by a high degree of subjectivity: by nature, they are nebulous, ill-defined and contingent upon the contributions of as many stakeholders as possible. As modelled in the Fig. 6.4, these two elements (subjectivity and kinetics) apply equally to the difficulty in addressing wicked problems.

6.4 Sustainable Development as a Wicked Problem

Sustainable development is encumbered by the inherent complexity arising from the involvement of a network of different stakeholders and contributing worldviews. This makes for markedly difficult situations in tackling problems of this magnitude and complexity. For example, citing the Species Survival Commission of the World Conservation Union (IUCN), Holling (2001) notes that sustainability "either in a region or a species, depends on interactions among internal and external factors" (p. 390). The internal factors may include "social, political, ecological or economic" aspects, while external factors include "foreign debt, structural poverty, global environmental problems and social/political/economic conflicts" (Ibid.). What this means is that coinciding factors such as a specific area's physical geography, cultural makeup and economic activities all contribute to make for the uniqueness of that particular area. Because of this distinctiveness, regional sustainable development is a tool used so that sustainable systems function at the organizational level (Ring et al. 1999). Nijkamp et al. (1991) characterize regional sustainable development in terms of fulfilling two goals: "It should ensure for the regional population an acceptable level of welfare, which can be sustained in the future [and] it should not be in contrast with SD at a supra-regional level" (p. 3). The reason behind the latter goal is the recognition that regional sustainability is a subset of overall global sustainability and that each region has its own pathway to achieving sustainability (Nijkamp et al. 1991).

Another complicating factor is that many elements which constitute sustainability or sustainable development are themselves considered wicked problems. For example, Rayner (2006) considers climate change, stratospheric ozone depletion and persistent organic pollutants (POP) as wicked problems in themselves. In this way it meets the qualification by Rittel and Webber (1973) that every wicked problem is a symptom of another wicked problem (property 8). Furthermore, there is also the element of different viewpoints among participating stakeholders. In Europe, the precautionary principle (see above) is typically embraced, while in the United States the proportional principle is embraced. The latter principle means that any regulatory regime must balance the costs with benefits in any intervention (Rayner 2006). According to Ravetz (2006), the failure of society to successfully transform itself to sustainability could see the demise of the human species. In this way, this particular dilemma is indicative of the "one shot" element of a wicked problem (property 5). Further elaborations on the characteristic way sustainable development is a wicked problem are outlined in Table 6.1.

Characteristic of the current state of sustainability research is that it is "cluttered with a multitude of seemingly random interpretations and competing frameworks, without definable routes to implementation" (Frame and Brown 2008, p. 227). A similar sentiment has been expressed by Espinosa et al. (2008) in the difficulties experienced in implementing any holistic approach, especially owing to what Midgley and Reynolds (2004) call the "complexity and uncertainty of natural and social phenomena; multiple and often conflicting values; and political

Table 6.1 Characteristics of a wicked problem applicability to sustainable development

Properties of a wicked problem (Rittel and Webber 1973)	Characteristics applicable to sustainable development
1. There is no definitive formulation of a wicked problem.	It is not possible to identify an exhaustive inventory of all possible challenges, opportunities, and alternative solutions associated with SD because these will continuously emerge as the issue is explored. Similarly, the solution set is marked by a number of ill-defined parameters and variables such as "reduce harmful emissions".
2. Wicked problems have no stopping rule.	There is no definitive end point where SD can be declared to have been achieved. Factors which can be implicated in a cessation of SD activities are contingent upon arriving at a consensus, which may be impossible. There is also a kinetic element at work which may alter preferences or the problem definition over time. Issues such as atmospheric pollution require long term monitoring and can be complicated by unknown synergistic effects with other emergent emissions.
3. Solutions to wicked problems are not true-or false, but good-or-bad.	There is no optimal solution to SD since it is impossible to optimize all stakeholder objectives simultaneously. This is based on the notion that contending parties to achieving SD have divergent objectives set as their goals. For example, the trade-off between economic activity and preserving endangered species has differences of degree, not differences of an axiomatic nature.
4. There is no immediate and no ultimate test of a solution to a wicked problem.	The full consequences of any actions taken to improve progress towards SD cannot be known for some time, if ever. The complexity inherent to a wicked problem makes arriving at a test for a solution impossible because of the difficulty in assessing ill-defined parameters. For example, although the elimination of Freon usage has seen repairs in the ozone layer in the southern hemisphere, the point at which it will be deemed to be sufficiently or completely repaired is contentious.
5. Every solution to a wicked problem is a "one shot operation"; because there is no opportunity to learn by trial-and-error, every attempt counts significantly.	Many actions taken to improve progress towards SD cannot be undone and any actions, or inactions, may have potentially catastrophic consequences. Also, the stakes are so high (species survival) that there is no room for error. The time scale involved also precludes any experimentation. For example, remediation efforts at superfund sites cannot be reversed.
6. Wicked problems do not have an enumerable set of potential solutions, nor is there a well described set of permissible operations that may be incorporated into the plan.	There are nearly limitless possible courses of action to improve progress towards SD and it is not possible to establish criteria to test whether all possible alternatives have been identified and considered. This problem is particularly acute given the high speed at which technology is proliferated.

(continued)

Table 6.1 (continued)

Properties of a wicked problem (Rittel and Webber 1973)	Characteristics applicable to sustainable development
7. Every wicked problem is essentially unique.	A large part of the challenge of SD is not knowing early on which solutions may be appropriate and that solutions used in the past may not be appropriate in addressing future issues. The uniqueness of the problem should be identified in order to achieve consensus on the problem definition. The aim here is to avoid generalizations and sweeping proscriptions such as "reduce coal use". Also, it may not be possible for technology to solve all problems. For example, large scale oil spills in the Arctic present challenges which are unique.
8. Every wicked problem can be considered to be a symptom of another problem.	The many challenges associated with SD, such as climate change, curbing population growth, non-renewable resource depletion, poverty reduction, and improving access to quality education, are interrelated and interdependent. Identifying causal attributes may involve an unacceptable degree of complexity. The tendency to reduce problems to simpler levels may result in inadequate solution spaces. For example, endangered species are usually treated on an individual basis rather than as a systemic whole and the preference for saving one species rather than another is highly contingent upon anthropomorphic preferences.
9. The existence of a discrepancy representing a wicked problem can be explained in numerous ways. The choice of explanation determines the nature of the problem's resolution.	There is no definitive way to determine the correct explanation or to deal with conflicting evidence and opinions for many SD challenges such as climate change. Building a medical incinerator in an urban setting may be described as a necessity by public health officials, while the local residents may be recalcitrant with the location of incinerator even though they may recognize the necessity of its use.
10. The planner has no right to be wrong.	There is only one planet Earth and the consequences of actions can have sweeping ramifications. Because of the time, cost and scale of wicked problems, there is no room for error. For example, reducing carbon emissions requires a correct solution owing to the fact that carbon has a 500 year lifecycle in the atmosphere.

effects" (p. 638). This has not stopped some from suggesting a transdisciplinary platform be established in dealing with issues of sustainable development. By transdisciplanarity, Hadorn et al. (2006) suggest the breaking down of borders which normally separate classical disciplines within science and the integration and synthesizing of many different perspectives. They state, "one way to do this is by focusing more directly on the problems, rather than the particular intellectual tools and models used to solve them, and by ignoring arbitrary intellectual turf boundaries" (p. 120). How exactly it is possible to focus on wicked problems at the expense of "turf boundaries" is not particularly well elaborated, nor is there

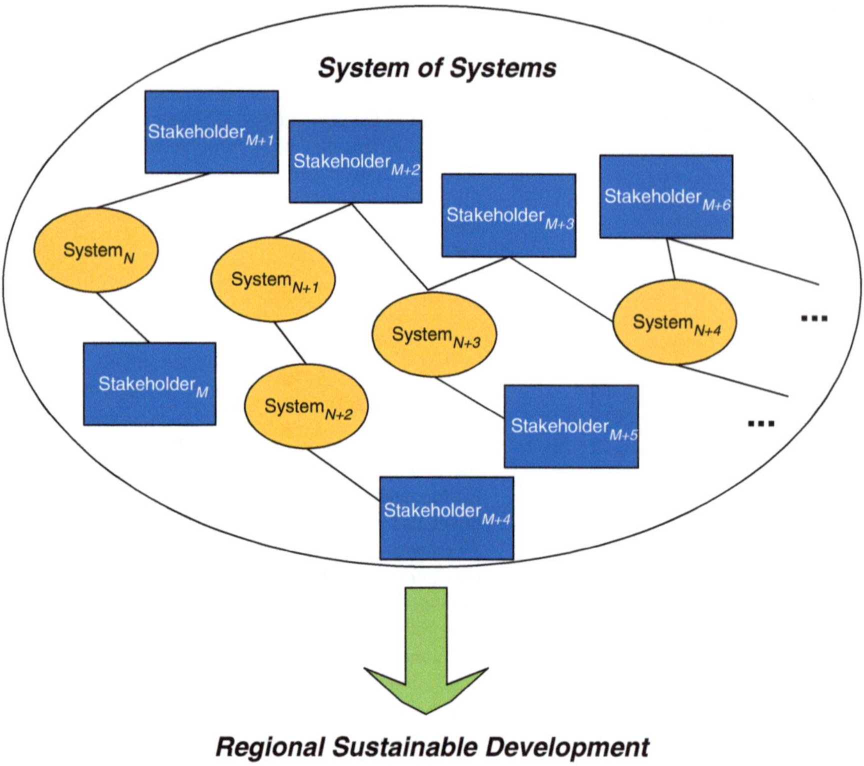

Fig. 6.5 Regional sustainable development and SoS

a well- established history to fall back on. For example, sustainable watershed studies typically are interdisciplinary by nature; however, many of these programs offered at the university level are subsets of traditional programs such as civil engineering. This problem is still ongoing with regard to many subjects in sustainable development.

Addressing sustainable development from a regional perspective involves a network of many different stakeholders (organizational, institutional, or otherwise) acting on behalf of multiple individual systems towards achieving overall regional sustainability (Fig. 6.5). Individual regional stakeholders could have a variety of different focussed tasks ranging from data gathering and dissemination, to decision-making and ground level implementation of regional sustainability actions. The data collected at the regional level could take on many different forms such as economic performance, biodiversity indicators or pollution monitoring. As such, there is a high degree of complexity and wickedness in such instances. This contributes to a situation which Jackson and Keys (1984) term the "systemic-pluralist contexts" of wicked problems (p.480), one which necessitates an overarching system to integrate all the individual systems, a so-called System of Systems (SoS). According to Meentemeyer et al. (2009), there has been growing interest in the last 10 years in the use of Systems of Systems (SoS) to "address large-scale, multidiscipline problems through the use of large complex systems containing multiple smaller systems" (p. 330). This arose out of the failures of traditional Systems Engineering (SE) to address complex problems, especially systemic issues which extend beyond the project level (Meentemeyer et al. 2009). Also, traditional SE has been faulted as not being well suited to handle ambiguity and uncertainty; as well, it de-emphasises the problem context and disavows the incomplete nature of solutions to wicked problems (Keating et al. 2003). The System of Systems Engineering Center of Excellence (in Meentemeyer et al. 2009, p. 331) claims that a SoS is "a complex purposeful whole that:

- is composed of complex, independent, self-organising component parts whose high levels of interoperability enable them to be composed into different configurations and even different SoS
- is characterised by contextual complexity that significant affects its behaviour and makes it difficult to understand
- has ambiguous and/or changing boundaries
- exhibits emergent properties."

Similarly, Keating (2005) claims the following characteristics of SoS (p. 2721), most of which compliment or augment the above specifications:

- A SoS involves the integration of multiple, potentially previously independent, systems into a higher level system (metasystem).
- The function of a SoS generates capabilities beyond what any of the constituent systems is independently capable of producing.
- Integration into a SoS evokes some degree of constraint for previously independent systems.
- A SoS brings together systems in order to perform a higher level mission/ purpose of which each member system plays an integral role.
- A SoS is a complex system and as such exhibits dynamic and emergent behaviour, is difficult to grasp, and problematic to engineer.

Although there still exist divergent views as to what exactly constitutes a SoS, there is some agreement concerning general principles, many of which are in accord with wicked problems and the demands of regional sustainable development. The need for SoS has come about by the recognition that complex or wicked problems must be addressed on a collaborative basis. Stated another way, Jackson and Keys (1984) assert that with regard to wicked problems, "There seems to be an unwarranted assumption that, once the pluralism [of a wicked problem] has been dissolved, the problems stemming from the systemic nature of the context will disappear as well" (p. 480). This is very reminiscent of property 8 of a wicked problem, namely that wicked problems should be dealt with at the highest possible hierarchical level. With regard to sustainability at the regional level, the piecemeal handling of sustainability issues is constrained by project level solutions to what is a dynamic and broad situation requiring a SoS. As noted by Kovacic et al. (2008), a SoS entails not only merely technical aspects to deal with wicked problems (i.e., individual systems), but also includes the social domain as well. Furthermore, the temptation to compartmentalize the elements which make up an overall regional sustainability platform diverges from the requisite indefinite composition of a wicked problem (property 1). There is also issue with the uniqueness of every wicked problem (property 7) because compartmentalization of sustainability tends to employ methods which overlook emergent behaviours (Keating et al. 2003). As such, at the regional level, sustainable efforts should not be appropriated by or relegated to one stakeholder—for example, a ministerial body. In cases which require a SoS, there is never one clear, single entity or body entrusted with decision making: "The multiple organisations managing the constituent systems must work together to make decisions and manage risks at a global level" (Meentemeyer et al. 2009, 342). It is under these sets of assumptions that SoS will be deemed purposeful in regional sustainable development.

6.5 Summary and Conclusions

This paper introduced the notion of regional sustainable development (SD) as a wicked problem. Sustainable development has a long and rich heritage marked by different philosophical approaches and principles. The debate about strong and weak sustainability is ongoing and has yet to show signs of reconciliation. The original propositions of Rittel and Webber (1973) regarding wicked problems were initially intended for public policy issues in the field of design. Nevertheless, Rittel & Webber's thinking in this matter has penetrated many diverse disciplines so that now there is some serious need for agreement on the applicability of this development. Many characteristics of wicked problems show overlapping and coherence, such that they may be said to exhibit two significant traits. First, they are dynamic and kinetic. Secondly, they are subjective not only because of the individual perspectives with regard to the many different stakeholders involved, but also due to the confluence of applied science with politics. Furthermore, it was demonstrated that SD exhibited each of the ten established characteristics of a wicked problem. Certain aspects of System of Systems (SoS) were also outlined and deliberated with regard to their applicability to address issues of sustainable development. Although in some facets this appears promising, more work in this area should be the subject of future research.

References

Adams WM (2006) The future of sustainability: re-thinking environment and development in the twenty-first century. IUCN, The World Conservation Union, Gland

Agyeman J, Evans T (2003) Toward just sustainability in urban communities: building equity rights with sustainable solutions. Ann Am Acad Pol Soc Sci 590:35–53

Allan L, Godfrey P, Allan GM (2008) Ports as sustainable complex systems. J Marit Res 5(3):55–66

Baker S, Kousis M, Richardson D, Young S (eds) (1997) The politics of sustainable development: theory, policy and practice within the European Union. Rutledge, London
Basiago AD (1995) Methods of defining 'sustainability'. Sustain Dev 3:109–119
Batie SS (2008) Wicked problems and applied economics. Am J Agric Econ 90(5):1176–1191
Berardini BV (2007) Complexity and command at the operational level of war. MECS 2007 conference proceedings October 2007. MECS, Norfolk, pp 38–47
Buchanan R (2009) Wicked problems in design thinking. In: Clark H, Brody D (eds) Design studies: a reader. Berg, Oxford
Churchman CW (1967) Wicked problems. Manag Sci 14(4):B141–B142
Conklin J (2006) Dialogue mapping: building understanding of wicked problems. Wiley, Chichester, UK
Christensen K (2009, Winter) Building shared understanding of wicked problems. Rotman Magazine, pp 16–20
Coyne R (2005) Wicked problems revisited. Design Stud 26:5–17
Espinosa A, Harnden R, Walker J (2008) A complexity approach to sustainability—Stafford Beer revisited. Eur J Oper Res 187:636–651
Frame B (2008) 'Wicked', 'messy', and 'clumsy': long-term frameworks for sustainability. Environ Plan C Gov Policy 26:1113–1128
Frame B, Brown J (2008) Developing post-normal technologies for sustainability. Ecol Econ 65:225–241
Frey H, Yaneske P (2007) Visions of sustainability: cities and regions. Taylor & Francis, London
Funtowicz SO, Ravetz JR (1993) Science for the post-normal age. Futures 25(7):739–755
Goldsmith E, Allen R (1972) Blueprint for survival. Houghton Mifflin, Boston
Gray CG (1914) Rent under the assumption of exhaustibility. Q J Econ 28(3):466–489
Grove R (1990) Threatened islands, threatened Earth: early professional science and the historical origins of global environmental concerns. In: Angell DJR, Comer JD, Wilkinson MLN (eds) Sustaining Earth: response to the environmental threat. Macmillan, Basingstoke, pp 15–29
Haag D, Kaupenjohann M (2001) Parameters, prediction, post-normal science and the precautionary principle: a roadmap for modelling for decision-making. Ecol Model 144(1):45–60
Hadorn GH, Bradley D, Pohl C, Rist S, Wiesmann U (2006) Implications of transdisciplinarity for sustainability research. Ecol Econ 60:119–128
Hitch CJ (1969) The universities' commitment to contemporary problems. Manag Rev 11(4):4–6
Holling CS (2001) Understanding the complexity of economic, ecological and social systems. Ecosystems 4:390–405
Hotelling H (1931) The economics of exhaustible resources. J Polit Econ 39(2):137–175
Jabareen Y (2008) A new conceptual framework for sustainable development. Environ Dev Sustain 10:179–192
Jackson MC, Keys P (1984) Towards a system of systems methodologies. J Oper Res Soc 35(6):473–486
Keating C (2005) Research foundations for system of systems engineering. IEEE international conference on systems, man and cybernetics, Waikoloa, Hawaii, 10–12 Oct 2005, IEEE, Los Alamitos, CA, pp 2720–2725
Keating C, Rogers R, Unal R, Dryer D, Sousa-Poza A, Safford R, Peterson W, Rabadi G (2003) System of systems engineering. Eng Manag J 15(3):36–45
Kovacic S, Sousa-Poza A, Keating C (2008) Complex situations: an alternative approach for viewing a system of systems. In: IEEE international conference on system of systems, 2008. SoSE '08, Singapore, 2–4 June 2008, (IEEE, Los Alamitos, CA, pp 1–6
Kraft J (1960) RAND: arsenal for ideas. Harper's Mag 221:69–76
Meadows DH, Meadows DL, Randers J, Behrens WW (1972) The limits to growth. Universe Books, New York
Meentemeyer SM, Sauser B, Boardman J (2009) Analysing a system of systems characterisation to define system of systems engineering practices. Int J Syst Syst Eng 1(3):329–346
Midgley G, Reynolds M (2004) Systems/operational research and sustainable development: towards a new agenda. Sustain Dev 12(1):56–64

Mosselmans B (2004) Time and value in the history of political economy. Found Sci 10:325–345
Naess A (1977) Spinoza and ecology. Philosophia 7(1):45–54
Neumayer E (2010) Weak versus strong sustainability: exploring the limits of two opposing paradigms, 2nd edn. Edward Elgar, Cheltenham
Nie M (2003) Drivers in natural resource-based political conflict. Policy Sci 36:307–341
Nijkamp P, Laschuit P, Soeteman F (1991) Sustainable development in a regional system: serie research memoranda. Faculteit der Economische Wetenschappen en Econometrie, Vrije Universiteit, Amsterdam
O'Riordan T, Voisey H (1998) The political economy of the sustainability transition. In: O'Riordan T, Voisey H (eds) The transition to sustainability: the politics of Agenda 21 in Europe. Earthscan, London
Ott K (2003) The case for strong sustainability. In: Ott K, Thapa P (eds) Greifswald's environmental ethics. Steinbecker Verlag Ulrich Rose, Greifswald
Parr A (2009) Hijacking sustainability. The MIT Press, Cambridge
Perelman M (1987) Marx's crisis theory: scarcity, labor and finance. Praeger, New York
Poundstone W (1992) Prisoner's dilemma. Anchor, New York
Ravetz JR (2006) Post-normal science and the complexity of transitions towards sustainability. Ecol Complex 3:275–284
Rayner S (2006) Wicked problems: clumsy solutions—diagnoses and prescriptions for environmental ills. Lecture delivered at the Jack Beale Memorial Lecture on Global Environment, University of New South Wales, July 25, 2006. Retrieved December 13, 2010, From: http://www.sbs.ox.ac.uk/centres/insis/Documents/jackbealelecture.pdf
Ring I, Klauer B, Wätzold F (1999) Towards regional sustainability: the need for interdisciplinary and applied research. In: Ring I, Klauer B, Wätzold F, Månsson BÅ (eds) Regional sustainability: applied ecological economics bridging the gap between natural and social sciences. Physica-Verlag, Heidelberg, pp 3–16
Rith C, Dubberly H (2007) Why Horst W.J. Rittel matters. Design Issues 23(1):72–74
Rittel HWJ, Webber MM (1973) Dilemmas in a general theory of planning. Policy Sci 4:155–169
Sachs W (1993) Global ecology and the shadow of development. In: Sachs W (ed) Global ecology. A new arena of political conflict. Zed Books, London, pp 3–20
Saha D, Paterson RG (2008) Local government efforts to promote the "Three Es" of sustainable development: survey in medium to large cities in the United States. J Plan Educ Res 28:21–37
Saul JR (1992) Voltaire's bastards: the dictatorship of reason in the west. Penguin, Toronto
WCED (1987) Our common future. Oxford University Press, Oxford

Chapter 7
Complexity and Command at the Operational Level of War

Vince Berardini

An actor in a complex system controls almost nothing but influences almost everything.

– Scott Page.[1]

As soon as there is life, there is danger.

– Ralph Waldo Emerson

Abstract The operational level of war for the U.S. military lies between two boundaries: tactics and strategy. Similarly, complexity lies between two boundaries: equilibrium and chaos. The symmetry is perhaps coincidental, but it highlights an important point, that is, both share common attributes. This paper explores those and describes the military as a complex organization confronting complex adversaries operating in complex environments. If we assume the character of war constantly changes, as many in the military claim, it follows that the character of future war must also change. What models and tools will enable operational level practitioners better understand this co-evolutionary dynamic? Complexity research offers a new perspective.

In 2007, Jonathan Schaeffer et al. published an article in Science titled, Checkers Is Solved (2007). This may seem trivial, but the implications are staggering: Checkers has 500 billion-billion (5E+20) possible positions and it took him and his team18 years to arrive at the mathematical proof. By comparison, chess has 1E+120 possible game variations, and, unlike checkers, it's unlikely to be solved anytime soon. Both checkers and chess are finite games of perfect information, and they are highly complicated. Warfare can also be thought of as a game, that is, in

[1] Leonid Hurwicz Collegiate Professor of Complex Systems, Political Science, and Economics, The University of Michigan - Ann Arbor. External Faculty, The Santa Fe Institute.

V. Berardini (✉)
Vince Berardini, Senior Analyst, ROLAND & ASSOCIATES Corporation, Monetary, CA
e-mail: vince.berardini@rolands.com

S.F. Kovacic and A. Sousa-Poza (eds.), *Managing and Engineering in Complex Situations*, Topics in Safety, Risk, Reliability and Quality 21, DOI 10.1007/978-94-007-5515-4_7,

a game theoretic sense. But important, nontrivial distinctions exist. War is not a game. Information is not perfect, nor is it finite. More importantly, warfare is more than complicated—it's complex.

Complexity is a relatively young science (c.1980s), unlike Newtonian physics, which permeated Western culture for more than 150 years. However, traces of complexity inquiries began as early as the late eighteenth century with Adam Smith; to the nineteenth century with Charles Darwin; to the early twentieth century with Henri Poincaré; and to the mid-twentieth century with Alan Turing, John von Neumann, Murray Gell Mann, John Holland, Edward Lorenz, and countless others.

War and many social problems are contemporary realities for which complexity science is offering new thinking. This chapter limits its scope to complexity at the operational level relative to armed conflict, and from a U.S. military context.[2] After reviewing basic foundations of armed conflict, I explore how problems can emerge in complex operating environments (OE)[3] and the tools commanders use to understand and resolve them. I contrast this with the tools and concepts available from complexity science to show why military problem solving today is "muddling through."

7.1 Armed Conflict (War)

The Western perspective of war dates back to the ancient Greeks when Thucydides reflected on the origins, causes, and consequences of the Peloponnesian War.[4] Also in the West, war remains a tension between theory and practice. The theory rests, in part, on the Platonic ideal; i.e., projecting an idealized plan of action on the problem and then actualizing that plan through force of will. Military practitioners sequence actions to achieve particular ends. This is a means-to-ends relationship, where the ends (or endstate) are conceived first, and then appropriate means are developed to achieve desired goals (Jullien 2004).

Perhaps the most quoted military theorist in the West is Carl von Clausewitz, the Prussian theorist who developed a theory of war in the early 1800s. At its most basic level, he described war as "nothing but a duel on a larger scale." He goes on to note that; "War is thus an act of force to compel our enemy to do our will" (von Clausewitz et al. 1976). To many in the West, these ideas remain the quintessential framework for the military.[5] While scholars still debate the contemporary

[2] Hereafter referred to as "military."

[3] The military uses "operating environment" or OE to describe the "composite of conditions, circumstances, and influences that affect the employment of capabilities and bear on the decisions of commanders." Joint Publication (JP) 3-0, Joint Operations.

[4] It should be no surprise that the literature on war is voluminous. My brief treatment here is meant only for context, highlighting basic attributes of war viewed from a U.S. military perspective.

[5] Antulio Echevarria, Clausewitz and Contemporary War, notes that Clausewitz had two ways of presenting his theory of war. One was the search for objective knowledge, the scientific, or cause-and-effect relationships to get at the truth about war; the other was as an "organized corpus of scientific observations." (p. 3)

relevance of Clausewitzian theory, there is little debate concerning the character of war—fundamentally, the nature of war remains unchanged, that is, compelling an enemy through force of will, but the character of war, or the ways in which it is conducted changes frequently. For example, the physical space where uniformed combatants and machines decide disputes on a battlefield no longer pervades Western military culture.[6] While artifacts remain, contemporary practitioners like Brigadier General Rupert Smith, assert that war occurs "amongst the people" (2005).

7.1.1 Levels of War

The military recognizes three levels of war: strategic, operational, and tactical.[7] These are designed to distinguish the employment of military and nonmilitary sources of power to accomplish assigned military objectives in a geographic theater of operations (Vego 2000). The stratifications exist because different command echelons (levels) have different responsibilities, aligned to resources and sources of power to accomplish assigned policy goals and objectives.

U.S. Department of Defense Joint Publication (JP) 3–0 Joint Operations, notes the strategic level of war is that level in which a nation, or group of nations, determines national or multinational (alliance or coalition) strategic objectives and guidance and develops and uses national resources to achieve those objectives (2010). The dimensions of time and space at the strategic level are naturally quite large. By contrast, the tactical level of war focuses on battles and engagements, where battles last somewhat shorter than engagements and are constrained by space and time. Typically, the tactical level includes a variety of actions that occur between belligerents who employ a variety of tactics, techniques, and procedures.

Joint Publication JP 3-0 also notes the operational level is one that links tactical employment of forces to national and military strategic objectives. The operational-level therefore lies "in-between" the tactical and strategic levels. It focuses on the design and conduct of strategies, campaigns, and major operations, as well as the organization and employment of military forces. The operational level of war emerged in the early twentieth century for a number of reasons: senior military commanders found it increasingly difficult to achieve strategic objectives by fighting single, or multiple decisive battles; science and Western industrialization changed the relationship between war and society in terms of how war was prepared for, conducted, and sustained; and finally, the steady expansion of the battlefield (Vego 2000). The operational level of war formally entered U.S. military lexicons in 1982.

---

[6] While it is generally true that military practitioners today no longer conceptualize war this way, problem-solving tools within the institution are based on canonical decision theories and methodologies.

[7] For large countries like the US that deploy forces abroad, or project power away from the continental U.S., the strategic level is sub-divided into national-strategic and theater-strategic levels.

The operational level remains a distinctive military feature, even though current doctrine is beginning to note limitations. For example, the wars in Iraq and Afghanistan over the past decade raised questions and criticisms from a number of senior officials: "The categories of warfare are blurring and no longer fit into neat, tidy boxes. One can expect to see more tools and tactics of destruction—from the sophisticated to the unique—being employed simultaneously in hybrid and more complex forms of warfare" (Gates 2009). If the categories of warfare have indeed changed, the co-evolutionary dynamic may signal an irrelevance of the levels of war. Political objectives designed at the strategic level and the physical interventions at the tactical level may no longer be sufficient to resolve complex national security problems.

7.1.2 Military Command and Control

Like armed conflict, the literature on military command and control (C2) is extensive, much of which includes an assortment of elements and principles; including theories, concepts, forms, functions, guidelines, relationships, processes, and even its own vernacular. These comprise the C2 system and what "the commander" must be, and be able to accomplish. Accordingly, Joint Publication (JP) 1–02, notes (2003):

> Command is the authority that a commander in military service lawfully exercises over subordinates by virtue of rank or assignment. Command includes the authority and responsibility for effectively using available resources for planning the employment of, organizing, directing, coordinating, and controlling military forces for the accomplishment of assigned missions. It also includes responsibility for health, welfare, morale, and discipline of assigned personnel (2003).

Military C2 clearly serves a foundational role for all service components. Interestingly, however, "control" remains an appendage to "command." This may be an historical artifact, but it nonetheless conflicts with the emerging science of complexity and contemporary realities.

7.2 A Complex Military Ecosystem

The operational-level environment where military interventions occur is complex. This environment, or state space, comprises diverse agents, groups, and organizations that connect and adapt to other agents in the environment. But if contemporary military interventions are complex, how about 237 years ago when the U.S. Army was first established? I argue that military interventions 200 years ago were complex, but less so than today. And yet, future interventions may not be.

The table below helps explain why. It illustrates a partial representation of the dimensions and categories of an open military ecosystem, and shows why contemporary interventions are complex.

	Military ecosystem		
	Dimension	Categories	Number of sets
Natural	Geography	Land, air, sea, space	4
	Time	Present, future	2
	Weather	Hot, cold, rain, snow (many more variants)	4
Social	Military Components	Army, Navy, Marine Corps, Air Force, Coast Guard	5
	US Branches of government	Executive, Congressional, Judicial	3
	US Intel community (IC)	CIA, DOD (AFISRA, INSCOM, DIA, MCIA, NGA, NRO, NSA, ONI), DOE (OICI), Dept of Homeland Security (I&A, CGI), DOJ (FBI, DEA), DOS (INR), and DOT (TFI)	16
	US Allies e.g. in Afghanistan	A total of 56 nations help support efforts in Afghanistan. Of those, 24 provide troops.	24
	International organizations	UN, IMF, EU, WTO, WHO	5
	Other actors	Taliban, Al Qaeda, Sunni, Shia,	4
Manmade	Warfighting functions	Intel, maneuver, fires, protection, sustainment, C2	6
	Types of Joint operations	Counterinsurgency, stability, peace, and ten others	13
	Military equipment	Tanks, ships, airplanes, artillery, rifles, pistols, trucks etc.	7

Suppose, for example, the military develops a plan to intervene in a crisis. For the sake of simplicity, suppose further the intervention is a counterinsurgency operation. The difficulty, of course, is determining which dimensions they should examine.[8] If we take a subset of the above, let's say we want to limit our examination to four or five dimensions, respectively. That means there are almost 500 ways to choose four dimensions (12C4=495), and nearly 800 ways to choose five dimensions (12C5=792). But contemporary military environments are much more diverse than this simple example could show.[9] For instance, each of the five service components consists of Active and Reserve force structures comprised of commissioned officers,

[8] The ability to decide which dimensions and categories might be relevant to the problem can be overwhelming. But that's only part of the challenge. The other part is which of these many categories are the ones that matter and which ones can be ignored? This is the frame problem. Essentially which categories are relevant to the operation requires us to associate it with some set of comparable categories. But determining which categories are comparable depends on knowing which categories are relevant (Watts 2011, p. 44).

[9] As we'll see in the next section, the Military Decision Making Process (MDMP) guides military planners to develop three courses of action for the commander and the mission. This is a bit ironic, considering the number of combinations that exist. Importantly, it highlights the shortcomings to traditional methods of coping with complexity.

warrant officers, noncommissioned officers, civil servants, enlisted and contractor personnel. Functionally, each component conducts, to varying degrees, intelligence, maneuver, fires, protection, sustainment, and command and control.

Second, these diverse structures and functions are connected in a hierarchical command and control system, where Soldiers, Sailors, Marines, and Airmen interact in physical space and in time. They are also connected virtually through information communication technology (ICT) that exists internal and external to their team, group, or command. Third, interdependencies exist between and within these diverse and connected structures. Typically, interdependencies occur between agents because their strategies, payoffs, and artifacts interact in meaningful ways.[10] Finally, agents and the organizations to which they belong adapt and learn. This assumes actors learn through interactions, which occur locally and/or virtually.

In the past, military structures and functionalities were less diverse, connected, and interdependent. While the Continental Army certainly adapted and learned, they were much less diverse, connected, and interdependent from the contemporary Army today. In the future, it is unclear how much more diverse, connected, and interdependent the military can become. There may come a point where complexity attributes reach a critical state. In other words, the system will equilibrate, essentially becoming a tangled mess; a system incapable of additional diversity, connectedness, interdependencies and adaptations.

7.3 The Military Problem Space

Sixty years ago the threat of war existed between the Soviet Union and the Warsaw Pact, and NATO and the West. The prevailing idea at that time was the political objective was achieved once the military's strategic objective was achieved—unconditional surrender. But that never happened.

The Cold War paradigm is gone, the landscape changed. Today there are no clear ends to achieve, only conditions in which political objectives can be achieved with other means. However, military force structure, its doctrine, processes, and political-military cultures remain wedded to concepts and planning methods designed for interstate industrial warfare modalities. But the experiences in Iraq and Afghanistan have changed this thinking, slowly. Meanwhile, commanders at the operational level remain immersed in a conceptual space of planning and problem solving that have no clear ends.

[10] In the past, social and economic theories assumed rational choice guided individual behavior and decision-making. In effect, individual actors were treated as homogeneous, all acting rationally. Those theories no longer hold because agents are heterogeneous, employing various psychological biases or following "rules" or heuristics. See also Thomas Schelling (2006, p. 17).

Planning military campaigns and major operations are outlined in joint doctrine, Joint Publication (JP) 5–0, Joint Operation Planning. Its foundational logic is the Military Decision Making Process (MDMP), which follows a systematic process: receive the mission, conduct mission analysis, develop courses of action, analyze and wargame courses of action, compare courses of action, approve course of action, and develop a plan or order.[11] In broad terms, commanders receive a mission statement from their higher-level headquarters; including specified tasks, and purpose. Mission analysis is a process of understanding problems, which then drives of the remaining planning steps. For problems at the tactical level, MDMP is useful because problems tend to be more tangible. At the operational and strategic levels, however, the process becomes only a guide. In essence, the structure and logic of the process cannot capture system complexities. As a result, practitioners often cite insufficient intelligence or vague strategic goals as the problem. While this can be true, MDMP remains less effective for operational-level problems.[12]

A simple model helps describe how military problems can emerge. In Fig. 7.1 the horizontal axis represents time and the vertical axis represents so-called threats to national interests. Many assumptions must hold for the model, one being the greater the problem the greater the threat to national interests. As the problem forms and emerges (a notional position below the x axis), agents in the system "take notice." If the problem continues to grow, a critical threshold may be reached where "tips" can occur, creating decision points for political leaders and military commanders to intervene.[13]

Long before formalized strategic directives are issued to intervene, military commanders have assessed the problem, the environment, and many other variables. If interventions are pursued, the assessment process culminates with an operations order (OPORD). The tools with which commanders and staff planners assess the environment and produce the OPORD are based on MDMP.

[11] Planning, in the military context, is derived from decision theory. That is, gather sufficient information, analyze the information, develop a number of choices, and make a decision. There are two conditions assumed in decision theory. First, if a decision problem is to have meaning, the choices must have known consequences; and second, a choice is meaningful only if the chooser has preferences. We quickly see why decision theory fails for the military practitioner. Often the commander does not know what the consequences will be (beyond a level established by his experience), and frequently the military commander is constrained by the directives given to him by higher authorities.

[12] Over the past several years, joint doctrine assimilated an enabling process for MDMP. Operational Design (OD), or "Design" is now a formalized method to aid commanders and staffs understand complex conditions in contemporary operating environments. See JP 5-0, Joint Operation Planning (2011), Chapter III, Operational Art and Operational Design.

[13] British humanitarian Conor Foley (2008) notes that interventions almost never succeed. He likens it to performing heart surgery using scotch tape… countries like hearts are complex adaptive systems. In complex adaptive systems, interventions must be approached differently.

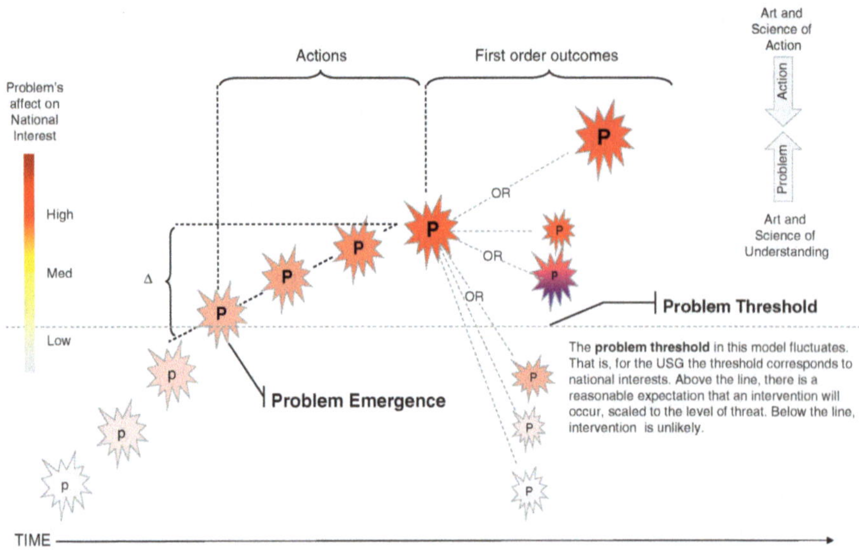

Fig. 7.1 Simple model displaying how military problems emerge

By the mid-1990s, joint doctrine hadn't formally recognized the existence of complex problems.[14] At the institutional level, complexity was treated as an outlier, a problem to be controlled and solved using traditional military decision-making processes (MDMP). While acknowledging the existence of complexity, the military institution remains reluctant to move beyond traditional processes. Even with enough time, enough information, and enough computing power, military problems remain elusive. What then are the options? As noted earlier "an actor in a complex system controls almost nothing but influences almost everything," the only near-term military option might be interventions that seek to manage complexity rather than attempting to control and solve it.

[14] Some branches of the military recognized complexity sooner than others. For example, the USMC began assessing non-linear dynamics and complex systems by the mid-1990s. See A. Ilachinski. Land Warfare and Complexity, Part II: An Assessment of the Applicability of Nonlinear Dynamics and Complex Systems Theory to the Study of LandWarfare. Technical report, Center for Naval Analyses, 1996. It wasn't until much later that military published, "Design for Operational Planning: A Primer for Joint Warfighters." United States Joint Forces Command (JFCOM), Joint Warfighting Center (JWFC) Joint Doctrine Series, Pamphlet 10, September 20, 2010.

7.4 In Search of Solutions

The use of simple, rugged, and coupled "landscapes" will be used as a metaphor and framework to help explain complexity and why some problems can be solved optimally and why others cannot.[15] Landscapes form because people, groups, and organizations all encode problems differently, that is, how a problem is represented determines how simple or difficult solutions will be. These are perspectives. The focus here will be on coupled landscapes. But first we need to understand how the other two operate.

Optimization problems generally fall into the simple landscape framework. For example, if we want to find the best price for a car or a computer, then classic decision methods work well. In other words, these problems can be optimized when choices don't interact with other interdependent, adapting entities. The method is familiar: gather available data, conduct an analysis, choose among a number of options, determine the payoff for each option, and then make a rational decision. Mathematically, the simple landscape can be depicted in a two-dimensional graph, the x-axis representing the range of outcomes (choices) and the y-axis representing the best payoff. Once graphed, the plot depicts a single peak, with the highest point being the optimal solution.

Rugged landscapes differ from simple landscapes because, just as the name implies, there are many local optima. Think of any rolling or rugged landscape (e.g., Colorado's Rocky Mountains, or New York's Shawangunk Mountains) and the metaphor becomes clear. The challenge of finding an optimal solution in a rugged landscape is that the number of possible choices can be combinatorially large, and those choices (of a single entity) interact; hence, the creation of local peaks. What makes the landscape rugged is not the number of choices per se; it is when those choices interact. To demonstrate how the landscape becomes combinatorially large, consider the case where the military desires a new ground combat vehicle (GCV). Clearly, a number of choices must be made. For example, will it be tracked or wheeled? How tall should it be? How thick should the armor be? What weapon's configurations should it have? And so on. Suppose 25 choices are available, and suppose the military program manager must decide on just five of them. The number of choices available is greater than 53,130! In other words, there are more than 53,000 ways to choose five features among the 25 available.[16] To see how 53,000 choices interact (a very large landscape), consider when the program manager decides to increase the thickness of the armor for protection. This decision increases the weight of the overall vehicle, which then requires a bigger engine, which requires more fuel; ultimately, military aircraft can no longer transport the vehicle because its too heavy. In rugged landscapes, more interactions mean greater ruggedness.

[15] Page (2009) uses the terms "simple," "rugged," and "dancing" landscapes. Dancing landscapes and coupled landscapes are essentially similar, that is, when one peak changes it affects the properties of the other.

[16] 25C5 is 53,130 possible choices.

Page (2009) notes that curing cancer or designing the Atomic bomb are extremely difficult problems, but as difficult as they are, they're not complex—e.g., a bomb is a bomb. The atom bomb became complex only after it was dropped. With enough resources and enough time to explore the landscape, problems like these can be resolved. One implication of rugged landscape problems concerns resource allocation, where designers and planners can reasonably assume return on investment. Just like the A-bomb, a new ground combat vehicle is a difficult problem; there are many local peaks (solutions). However, getting from rugged landscapes to coupled landscapes requires interdependences between our actions and the actions of other entities that learn and adapt.

Earlier we noted that commanders and the operational level often intervene within a complex ecosystem. This is because diverse actions and behaviors in one system are interdependent with the actions and behaviors of entities in other systems. Consider the example from the previous section in which a commander at the operational level receives a mission to defeat an insurgent force. To accomplish the mission, operational-level commanders design campaigns and major operations—i.e., bring together strategy and tactics to accomplish national policy goals and objectives. Suppose the commander assembles a team of adept military theorists, strategists, and tacticians who work for days coming up with the best possible military intervention. Suppose further that on the day the mission will be executed (D-Day), insurgents conduct a surprise attack first. What happened? The problem these planners were solving for (defeating the insurgency under the existing plan) no longer exists. It's gone; the landscape just shifted. What was once a local peak is now a small hilltop. In the military, a common aphorism is a plan seldom "survives contact with the enemy."

7.5 The Limits to Problem Solving: Harnessing Complexity

So far, we have examined armed conflict, levels of war, and military command and control—a brief tour of the battlefield so to speak. A fair amount of this chapter also focused on operating environments and a type of problem operational-level commanders might address. At this point, I want to examine the frameworks contemporary scholars use to understand and manage complex problems. Before doing that, it is important to examine contemporary military literature for the ways in which complexity is addressed.

The two military publications are US, Department of the Army. FM 5-0 (2010), The Operations Process; and U.S. Joint Forces Command's, Joint Warfighting Center (JWFC) Publication 10 (2010), Design in Military Operations: A Primer for Joint Warfighters. Both documents address complexity tangentially through an organizational methodology called "Design." FM 5–0 describes Design "[as] a methodology for applying critical and creative thinking to understand, visualize, and describe complex, ill-structured problems and develop approaches to solve them" (2010, 3–1). The JWFC Pub 10 (2010) describes Design "[as] a repeatable methodology of reasoning that helps commanders understand how to change a complex adaptive system from 'what is now' to 'what is better'" (2010, 3).

Both FM 5–0 and the JWFC Pub 10 recognize the importance of complexity. However, their descriptions differ in one significant way: the Army's Field Manual seeks to "develop approaches to solve [emphasis mine] complex, ill-structured problems," whereas the Joint Forces Command Publication seeks to "change [emphasis mine] a complex adaptive system from 'what is now' to 'what is better.'" I argue that JWFC Pub 10 comes closer to contemporary research in complexity because it acknowledges a fundamental principle, that is, controlling a complex systems through interventions is illusory; hence, it will unlikely be solved. Indeed, recent history is full of attempts to control and solve complex problems.[17]

If problems in simple landscapes can be solved and optimized, and if problems in rugged landscapes can also be solved (given enough time and resources), how should organizations approach problems emerging in coupled landscapes? According to complexity science, the best we can do is to learn how to harness and manage complex adaptive systems.

Page (2009) notes that we must first recognize when a situation is complex—not all situations are. Once recognized, the attributes of diversity, interdependence, connectedness and adaptability can be "dialed up" or down as the situation warrants. For example, system diversity is generally a desired attribute because without sources of diversity (through selection), the system can become too homogenous and it approaches a critical state. Diversity also prevents errors. Nonetheless, if we ramp up diversity too much the behavior of the system becomes chaotic or extremely sensitive to initial conditions (ESTIC). Next, observing the level of interdependence is important because if the actions and behaviors of agents in the system become too interdependent, the system can again be pushed to a critical state—this happens when organizations push for higher levels of efficiency. For connectedness, we need to seek out potentially synergistic links and cut out those that limit responsiveness. Finally, adaptation is concerned not with increasing the speed at which agents adapt, but increasing their learning; therefore, it is important to observe how the system defines its goals and objectives.

Robert Axelrod and Michael Cohen (1999) contend that a framework is needed for thinking through complex settings,[18] which allows new questions and new possibilities for intervention. Each of these concepts is relevant to new situations, and so we must ask how each of them can be interpreted. Axelrod and Cohen focus on three key concepts: agents, strategies, and populations. In complex systems, agents in a population are not the same. They exist within a larger population who share some detectable combination of features, distinguishable by properties or behaviors that are observable by others entities in the population or by outside observers.

[17] Iraq and Afghanistan are two examples. But outside the military, the national economy, crime, and climate change are just a few problems that resist solutions as well.

[18] That framework consists of 12 major concepts: strategy, artifact, agent, population, system, type, variety, interaction pattern, space (physical), space (conceptual) selection, and success criterion or performance measure.

Agents also have a number of properties: location, capabilities, and memory. The way an agent responds to its surroundings and pursues deliberate goals describes the concept of strategy. Here, the essential feature to analyze is how agent strategies change over time through such processes as reproducing and copying. Finally, if we want to harness complexity, then analyzing a population of agents is important for three reasons: as “a source of possibilities to learn from, as recipients for a new-found improvement, and as a part of your environment” (2000, 5).

A third way of exploring with the hope of harnessing complexity is Agent-Based Modeling. Before explaining its contribution, we should note that, in general, scientific knowledge advances two ways. The first is by employing mathematics and axiomatic methods to arrive at proofs. The advantage of mathematics is its logical consistency. One drawback, however, is its starkness. Another way to advance scientific knowledge is through narratives. Narratives are flexible, but they can be logically inconsistent. A third way, is Agent-Based Modeling, or a computational approach. This approach is logically consistent and flexible, it allows for exploration, and the modeler can run counterfactuals. As a primary means for exploring complexity, computational models are naturally suited to these problems because “they easily embrace systems characterized by dynamics, heterogeneity, and interacting components” (Miller and Page 2007). The computational approach is relatively new, but it opens up new frontiers for exploration for the military.

7.6 Summary and Implications

Classical decision theories, rational choice methodologies, and traditional systems analysis approaches are ineffective in complex systems; they are not the right tools for commanders at the operational-level. And yet, the military remains reluctant (with exceptions) to embrace the idea that operational-levels problems cannot be solved. Why? In general, because it goes against the prevailing culture; one that operates best with clear ends, ways, and means. If the culture persists, however, seeking control, order, and solutions in complex operating environments, then progress will suffer. Like it or not, complexity is here to stay.

Private industry, academia, and government-funded think tanks have explored complexity since the early 1980s. Prior to that time, game theory (Von Neumann and Morgenstern) general systems theory (Bertalanffy, Rapoport, Boulding), and chaos theory (Poincare, Lorenz, Feigenbaum, Mandelbrot) provided frameworks for thinking about previously unexplored and unexplained phenomena. Complex adaptive systems theory evolved from these earlier works and continues to evolve: new frontiers in complex responsive processes (Stacey 2007; Simpson 2007), variation-selection-interaction (Axelrod and Cohen 1999), agent-based modeling (Miller and Page 2007), and network science (Watts 2011). While the military evolved along with these developments, it did so cautiously. This must change.

Today is not the environment of 10, 20, or 30 years ago. Clearly, contributions made then remain invaluable, but ideas are not immutable. There was a time between the 1600 and 1800s when wise people thought the Newtonian paradigm was unassailable. Today, senior military leaders are concerned about the complexity they encounter. Robert Axelrod and Michael Cohen noted, "We all must intervene in Complex Adaptive Systems daily. We all face situations where the classical approach of formulating alternative actions and their likely consequences assumes more understanding and predictive power than we actually have" (1999, p. 160).

Commanders at the operational level of war know complexity; they live it daily. Understanding complexity better may allow them to understand and plan interventions that harness and shape complex phenomena, rather than relying on the traditional means. More importantly, military problems are beyond complicated; they reside in complex ecosystems, which require better understanding of the geopolitical conditions through complexity and network science. We can hope this will enable civilian and military leaders alike to pursue options other than war.

References

Axelrod RM, Cohen MD (1999) Harnessing complexity: organizational implications of a scientific frontier. Free Press, New York

Foley C (2008) The thin blue line: how humanitarianism went to war. Verso, London

Gates RM (2009) A balanced strategy: reprogramming the pentagon for a new age. Foreign Aff 88(1):28–40

Joint Publication (JP) 1-02, Department of Defense Dictionary of Military and Associated Terms, 12 Apr 2001 (as amended through 5 June 2003)

Joint Publication 3-0, Joint Operations (September 2006, w/Update 2010)

Joint Publication (JP) 5-0, Joint Operation Planning (2011)

Jullien F (2004) A treatise on efficacy: between Western and Chinese thinking. University of Hawaii Press, Hawaii

Miller JH, Page SE (2007) Complex adaptive systems: an introduction to computational models of social life. Princeton studies in complexity. Princeton University Press, Princeton

Page SE (2007) The difference: how the power of diversity creates better groups, firms, schools, and societies. Princeton University Press, Princeton

Page SE (2009) Understanding complexity. The Teaching Company, Chantilly

Schaeffer J, Burch N, Bjornsson Y, Kishimoto A, Muller M, Lake R, Lu P, Sutphen S (2007) Checkers is solved. Science 317(5844):1518–1522. doi:10.1126/science.1144079

Schelling TC (2006) Micromotives and macrobehavior. Norton, New York

Simpson P (2007) Organizing in the mist: a case study in leadership and complexity. Leadersh Organ Dev J 28(5):465–482

Smith R (2005) The utility of force: the art of war in the modern world. Penguin Group, New York

Stacey R (2007) The challenge of human interdependence: consequences for thinking about the day to day practice of management in organizations. Eur Bus Rev 19(4):292–302

US, Department of the Army. FM 5-0, The Operations Process. Washington, DC: Government Printing Office. March 2010

US Joint Forces Command, The Joint Warfighting Center Joint Doctrine Series, Pamphlet 10, Design in Military Operations: A Primer for Joint Warfighters, September 2010

Vego MN (2000) Operational warfare. U.S. Naval War College, Newport

von Clausewitz C, Howard M, Paret P (1976) On war. Princeton University Press, Princeton

Watts DJ (2011) Everything is obvious: once you know the answer. Crown Business, New York

Chapter 8
The Wicked Problem of a Cognitive Environment in Complex Situations

Ryland C. Gaskins III

Abstract Complexity in the cognitive environment can have a significant impact on the success or failure of complex situations. The wicked problem lies inherently in the complexities of human and social dynamics that are intrinsic to so many complex situations especially likely to involve ambiguity and lack of clarity for decision making as well as uncertainty of situational understanding in a dynamic cognitive environment. Understanding the cognitive environment offers a heightened level of situational understanding imperative in a complex and dynamic situation where pattern recognition and problem solving are a challenge and the cost of mistakes is high. This paper offers strategies and examples for understanding and moving toward mastering the wicked problem of the cognitive environment.

Keywords Cognitive • Wicked • Complexity • Analysis • Knowledge structure • Mental map

8.1 Introduction

Cognitive psychology followed a period of behavioral based research when it was realized that behavioral theories were not sufficient to explain underlying processes in memory and learning. Since, cognitive psychology has gone through significant changes over the past 30 years going through a birth and more recently, a rebirth. Initial focus in cognitive psychology was on memory and learning and produced findings relating to intellectual abilities. This opened up the question of how individuals put together information for their understanding of how that level

R.C. Gaskins III, Ph.D. (✉)
Human Factors Modeling and Simulation,
234 High Water Ln, Kilmarnock, VA 22482
e-mail: ryangaskins@gmail.com

S.F. Kovacic and A. Sousa-Poza (eds.), *Managing and Engineering in Complex Situations*, Topics in Safety, Risk, Reliability and Quality 21, DOI 10.1007/978-94-007-5515-4_8,

of cognitive ability was achieved. More recent cognitive psychology has expanded this question to fully understand individual, team, group and cultural processes involved in establishing goals, judgments, and decisions. Klein (2006) recognizes the changing focus in terms of "framing the problem" with regard to goal change. We are now realizing that the cognitive environment is indeed a wicked problem because of the dynamic nature, lack of awareness with unclear goals, and vastly conflicting values and perceptions. Additionally, the issues of social cognition contribute to the wicked nature of the cognitive environment problem.

Social cognition also influences mental representation but with additional focus on the underpinnings of social inference. Gasson (2005) examined social cognition from a case study in the context terms of: (1) socially-situated cognition (interpretive framing process), (2) socially-shared cognition (joint framing), and (3) distributed cognition (overlapping frames). Interactions were observed as measured from a boundary-spanning perspective between the three levels on consensus-building. This shows the complexity of reciprocal influences of behavior, surroundings and person/cognitive factors stemming from Bandura's (2000) social cognitive theory.

Viewing the cognitive environment from the perspective of "mastering a wicked problem" means we need to adapt by developing measures and criteria that focus on these underlying mental models of abstract concepts for individuals and groups. Such an approach is an effort to focus the attention on the most relevant and useful cognitive environment variables in complex situations..

Three studies are presented with emphasis on the cognitive elements as they relate to mastering the wicked problem of the cognitive environment.

8.2 Study 1: Generalizing Human Error Rates from Air Traffic Control Operator to Nuclear Power Plant Operator

In a study to generalize human error rates from one complex environment to another Fleishman's Ability Requirements Taxonomy (Fleishman and Quaintance 1984) and the Task Characteristics Taxonomy (Farina et al. 1990) was applied (Gaskins 1995). The Task Characteristics Taxonomy includes factors of goal, response, procedure, stimulus and stimulus–response relationship which gets at the cognitive element of procedural decision making and simultaneity of response. Similar measures were obtained indirectly from other dimensions such as degree of operator control, difficulty of goal attainment, feedback, number of procedural steps, and variability of stimulus location.

The study also examined reliability levels with an intervention of advance organizers as a means for organizing information. Of relevance to the present paper are two results. The cognitive dimension of simultaneity of response had the strongest reliabilities and the greatest prediction of human error in the advance organizer condition within a multiple regression analysis of variance context.

The results support the conclusion that training using either advance organizers alone or in combination with traditional techniques do indeed improve rater reliability and evident understanding of the complex situation.

8.3 Study 2: Team Communications in Submarine Operations

The cognitive environment was also the focus of a study examining submarine crew communications (Gaskins 2006). The method employed three types of analysis: Systematic Human Error Reduction and Prediction Approach (SHERPA) (Baber & Stanton 2004), Link Analysis, and Task Communication Analysis (TCA) (Klein 1997).

The SHERPA analysis revealed goals, operations and plans to be identified in an hierarchical manner and coded for task classification (action, retrieval, checking, selection, and information communication), task likelihood of human error and consequences of human error. Link analysis was used to analyze the flow, means and type of communication. Task communication analysis consists of task analysis and breakdown into component tasks and steps, personnel analysis, and communication analysis based upon the previous two analyses. The communication is analyzed in terms of purpose of the communication and which activity is being coordinated, the sender/receiver, closed or open loop, general content, criticality, and hindrances to communication.

Results demonstrated the relevance of the wicked cognitive environment in a complex situation. Cognitions that led to human error were identified in a variety of submarine operational evolutions involving teams of participants in the communication processes.

8.4 Study 3: Crisis Action Planning and Complex Adaptive Decision Making with Gaming and Mental Mapping

In a study by Gaskins et al. (2005), crisis situations were created in a gaming environment in combination with Mental Mapping techniques. Crisis situations are often complex in nature, involving a great deal of disparate information and heavy time constraints. In the limited time available for a crisis, the commander and staff cannot collect all information on everything. In conjunction with time constraint considerations, the commander and planning staff are often faced with complex situations that span across the PMESII (Political, Military, Economic, Social, Infrastructure, and Intelligence) elements. The abundance of information with time constraints requires the commander and staff to make trade-offs in terms of the amount of information and the diversity of the information across the PMESII. These different information gathering strategies can be perceived across two dimensions, breadth and depth of information. Breadth of information refers to the diversity of information across the domain whereas depth of information focuses on drilling down in a particular area to obtain detailed information in that topic area. This study investigated how various strategies for knowledge building and information gathering influence crisis action planning (CAP).

The experimental setting used was a city planning scenario which shared many of the same characteristics as a military command and control situation. Primarily, a city planning situation requires information on a wide range of information sources and categories. The city planning scenario used involved six relevant and interrelated information categories to mirror the six PMESII elements and placed the participant in a crisis situation as the mayor of the city. A combination of Army and Navy ROTC students were used in conjunction with regular college students as participants.

Three experimental conditions were created to test the effect of different combinations of breadth and depth information gathering strategies for crisis assessment. The participants, as the acting mayor, had 60 min to assess the current situation with the information provided. Condition 1 represented depth of information by allowing the participant access to only one information category but a full 60 min to study it in depth. Condition 2 represented a balance of breadth and depth by allowing access to three information categories but only 20 min per category to examine the information. Condition 3 represented breadth of information by providing information on all six categories to the participant but only 10 min per category.

The participants were assessed on three aspects of planning: (1) understanding, (2) planning performance, and (3) predictive accuracy of the city conditions. The balanced condition displayed the highest scores for each of the areas of planning. Even though the lack of sample size reduced the statistical power in many cases to find significant differences between the conditions, the overall pattern of results strongly points toward a balanced approach across information breadth and depth. The implications of this finding suggest that the most effective strategy for a crisis action planning situation is a balance of both information breadth and depth. This could be accomplished by prioritizing the information categories according to the specific crisis situation at hand and then gathering as much information within those high priority and relevant information areas. This provides the flexibility the planners need to adapt their information gathering strategies to the current crisis situation. It also promotes awareness of the relevant PMESII elements while maintaining some degree of information detail to be collected and processed by the planning staff and commander. Additional research is needed to both validate these findings and to further expand the area of strategy use in crisis action planning situations.

8.5 Conclusions

Mastering the wicked problem of the cognitive environment is possible with careful attention to the myriad cognitive components and issues surrounding a complex situation. Three studies presented here exemplify how to quantify and qualify the cognitive environment with mental mapping techniques as a means for assessing or measuring change.

In study 1 it was the complexity of having to make simultaneous responses along with low operator control of the stimulus and number of elements per output unit that were most predictive of human error in the, highest inter-rater reliability groups of advance organizers or advance organizers plus traditional training groups. Study 2 demonstrates the cognitive level of team communication in a complex environment and relates it to human error probabilities. Study 3 presents a laboratory study using a game platform to create a complex situation where participants facing an election crisis. This study also demonstrates the use of mental mapping metrics as a means of assessing levels and changes in understanding. Study three illustrates the importance of balancing both breadth and depth of information for planning and understanding.

Future research and application should address this challenging and wicked cognitive environment. Addressing and effectively dealing with the cognitive issues is imperative for both current and projected understanding of a dynamic environment constrained by time.

References

Bandura A (2000) Social-cognitive theory. In: Al K (ed) Encyclopedia of psychology. American Psychological Association/Oxford University Press, Washington, DC/New York

Baber C, Stanton NA (2004) Task analysis for error identification. In: Diaper D, Stanton NA (eds) The handbook of task analysis for human-computer interaction. Lawrence Erlbaum Associates, Mahwah, pp 367–379

Farina AJ, Wheaton GR, Gaskins RC, Fleishman EA (1990) The task characteristics rating scales (Revised, Form C). Center for Behavioral and Cognitive Studies, George Mason University, Fairfax

Fleishman EA, Quaintance MK (1984) Taxonomies of human performance: the description of human tasks. Academic, Orlando

Gaskins R (1995) Training for increased rater reliability on the task characteristics taxonomy and Fleishman's ability requirements taxonomy using advance organizers. Doctoral Dissertation, George Mason University, Fairfax, VA

Gaskins RC, Boone CM, Spain RD, Mihalecz MC (2005) Knowledge – building strategies for crisis action planning, Joint Futures Laboratory, Concept Development Pathway, J9, Space & Decision Superiority, July 21, 2005

Gaskins RC (2006) Importance of modeling & simulation – an international view. In: Proceedings modeling & simulation (M&S) – a key technology for transformation, Kiel, Germany, 22–24 May 2006

Gasson S (2005) A soft systems analysis of social cognition in boundary-spanning innovation. System Sciences, HICSS, 2005. In: Proceedings of the 38th annual Hawaii international conference on systems sciences, Washington, DC

Klein G (1997) Applied cognitive task analysis. Klein Associates, Inc., Fairborn

Klein G (2006) Making people stupid. Presentation at the UHSI conference, Mystic, CT, June 7, 2006

Chapter 9
Reverse Decision Making: An Interpretive Framework for Pragmatic Decision Making

Samuel F. Kovacic

Abstract Decision making in Engineering Management suffers from Mitroff's Type III error; solving the wrong problem precisely, Mitroff (1998). Engineering Management requires a holistic interpretation of theory to provide for the diversity in decision making, often imposed by the pragmatic nature of the situation. As a multidisciplinary field Engineering Management challenges decision makers' ability to explain phenomena within the aggregate of each individual disciplinary boundary. Whether the failure is from the intractable nature of the individual disciplines that make-up Engineering Management or through the efforts of integrating misaligned perspectives generated from each discipline, Engineering Management suffers from the ensuing uncertainty and complexity that challenge decision makers. It is assumed that there is sufficient overlap between the two disciplines to overcome any integration issues, however, the gaps are generally obscured rather than addressed by the overlap.

This aggregation in disciplines leaves gaps in terms of making coherent decisions; each discipline is immersed within its own lexicon and axioms and is either subsumed or obviated entirely by the dominant discipline. These gaps become particularly poignant in wicked problems and are the focus of this paper, within these gaps lie uncertainty, and with it emergent and dynamic properties that over time constantly change the nature for how the problem is framed. Mainstream decision process follows a substantive approach that relies on ergodic and monotone conditions for the most effective decision choices. The central theme for this chapter is derived from the premise that a substantive decision approach is less than effective in dynamic situations where emergence and randomness are prevalent and a process approach would prove more utility.

S.F. Kovacic (✉)
Batten College of Engineering and Technology, Department of Engineering Management and Systems Engineering, Old Dominion University, Kaufman Hall 242, 23529, Norfolk, VA, USA
e-mail: samkovacic@gmail.com

S.F. Kovacic and A. Sousa-Poza (eds.), *Managing and Engineering in Complex Situations*, Topics in Safety, Risk, Reliability and Quality 21, DOI 10.1007/978-94-007-5515-4_9,

The general thesis for this chapter:

> *Execution of a decision process in a complex situation that sustains a continuous selection of plausible possibilities will avoid the need for probabilistic states as a catalyst for the decision maker*

This chapter intends to show how complex situations are antithetical to a substantive decision process and offers an alternative process approach towards wicked problems. This paper explores the nature of the multidisciplinary problem set and the challenges it poses on current decision processes and discusses the implications of decision making for addressing wicked problems. Finally the paper introduces the concept of an interpretive process for dealing with complex situations, Reverse Decision Making (RDM), for making decisions in complex situations. The intent of this chapter is to support a decision process that obviates the inherent complexity found in emergent dynamic environments for which multi-disciplines are created to address.

9.1 Preface

This paper is all about Reverse Decision Making (RDM), an interpretive framework for decision making in complex situations, it explains the why and what of decision making but falls short of discussing how a problem might be perceived for which a decision is made, it is for this reason a preface is provided; to highlight the criticality of bounding for RDM – or rather a theory for bounding relevant to complex situations and their problems, for it is bounding that enables how the situation is perceived.

RDM exists within philosophically bounded principles that define who and what we do, Sousa-Poza (2006):

- Reality exists both separate and part of the observer, and is beyond the observer's full understanding
  - Absolute statements of reality are presumptuous.
- Knowledge and knowledge development requires the bounding of [a part of] reality to extract a bounded domain
- The domain bounds all that is knowable, not necessarily known.

The paradigmatic congregation of these axioms create a situation where understanding exists in the whole yet can only be perceived by its parts – and yet the aggregation of the parts is not representative of the whole. To have it any other way would mean that the whole can never be or that the parts are all that can be. Conversely this suggests that parts are derived from the whole but the whole cannot be completely perceived by the derived parts. This requires a special kind of bounding construct that affects a certain forensic complexity,[1] allowing for both abductive,

[1] Forensic complexity implies the reduction within the whole while maintaining coherence with the irreducible nature of the whole.

inductive, or deductive reasoning for substantiating knowledge; accommodating for the existence of the known's, the known unknown's, and most importantly the Kantian unknown/unknown's. That which exists within the individual's perspective is that which is reducible and that the bounding mechanism resides in the irreducible whole, which becomes the function of the interaction of each perspective towards the holistic perspective; this is the formula for bounding within complex situations. This would suggest that because of the temporal nature of the perspectives conceptual bounding must be nearly atemporal by nature if it is to be of any value.

How do you create a boundary where specific boundaries are elusive? As a whole it is intuitively done, leaps-of-faith are made daily in every aspect of people's lives most prevalent in their choices, for which decision making is studied. Yet amidst all the examples or hypothetics that are mustered no empirical proof beyond this narration is offered, and yet a condition for which Reverse Decision Making cannot ignore.

In closing consider these rhetorical questions as metaphors for understanding the significance of bounding in complex situations:

- How does a group of people achieve such great deeds or misdeeds solely on their belief in the existence of God of which no proof exists?
- How does a family function cohesively beyond the interpretation of the technical definition of the structure of what a family is?
- How do disparate people rally into a nation to overcome insurmountable hurdles?

RDM accepts this paradigmatic theory through methodology that provides for methods and tools for situational bounding to occur. It is this that makes research into RDM significant; as understanding of the environment increases a philosophical methodology and process exists to exploit these insights. RDM allows for viewing a phenomenological problem coupled with its complementary nuomenological double, accepting the nuance that uncertainty is better accepted than extracted or obviated. These are the conditions a bounding theory within complex situations must account for, it is critical that our methods and techniques can affect the same.

9.2 Introduction

Decision making in Engineering Management suffers from Type III error, solving the wrong problem precisely. Engineering Management is a multidisciplinary field, ergo decision making in multidisciplinary fields are relegated to solving the wrong problem precisely. Engineering Management is greater than the sum of its parts (engineering and management) it is a holistic interpretation of differing disciplines aligned towards a common perspective for greater understanding. The conflicted uncertainty generated by this merger, whether from the integration of two disciplines or the paradigmatic tension generated from multiple perspectives, seems to hang like a shroud over the field of Engineering Management. Combining these two disparate disciplines, implying that a meta-type discipline would emerge, may have caused more uncertainty than certainty; the ambiguity is apparent when interpreting current academic curriculum for Engineering Management. A review of six U.S. universities[2] offering Engineering Management programs resulted in six thematically different descriptions,[3] reinforcing the notion of ambiguity in Engineering Management and whether it contributes towards generalizable knowledge, or, more importantly, how to transfer that knowledge for viable application in a pragmatic world.

It is not contended that Engineering Management deals with high levels of uncertainty and its implications such as emergence, non-ergodic, and non-monotonic conditions that lend itself to multiple interpretations, what is in contention is how the decision making methods for dealing with these conditions have not benefited by the integration of the disciplines. It also suggests that the dilemma that engineering management face is endemic to all multidisciplinary endeavors where two disparate disciplines are merged. Invariably techniques for solving multidisciplinary problems generally are entrenched in one discipline and made to "fit" in the other for a satisficing solution. A position that is less than helpful particularly with decision makers who must make decisions in indefeasible dynamic conditions for which Engineering Management seems to have accentuated rather than diminished.

9.3 Purpose

The purpose of this chapter is to posit an interpretive framework for decision making that obviates the inherent complexity found in emergent dynamic environments for which multi-disciplines, such as Engineering Management address.

The central focus of this paper is derived from a set of statements that exist for complex situations:

Statement 1: Sven Ove Hansson (1994) stated: "we do not decide continuously", add to this 'but we do decide in the continuum'. The decision process has been relegated to the whims of the substantive domain rather than an approach conducive to an environment rife with constant change.

Statement 2: The stages of decision processes are not congruent to the emergent nature of complex situations. Bound, analyze, and decide empower the decision maker into believing a succession of linear choices will produce a desired end result. This could become problematic if any one choice can change the outcome sufficiently that the end result is no longer desirable or achievable.

---

[2] A thematic analysis of the definition of engineering management from six U.S. colleges: Stanford, MIT, Berkley, Univ. of Missouri, Purdue, and Old Dominion Univ. was conducted by the author by extracting themes from the mission statements and curriculum of each university and compiling into common themes for comparison.

[3] Although the themes were not grossly different it was sufficient to generate ambiguity particularly in their meme's and highlight the separation between what skills and value-added each university was offering.

Statement 3: Substantive decision processes is a predictive process and yet decision making is a temporal response. A substantively primed decision process bounds the decision set with conditions not readily recalled, limiting any empirical or rational historical reflection for learning within the situation.

Statement 4: The decision process reduces irreducible problems down to alternatives that imply and impose a projected result that may not be viable or easily abstracted to the current problem. Once a decision space is reduced into bounded responses the act of bounding in complex situations can no longer show a measurable relationship back to the abstracted form. A process is needed that maintains a coherent relationship between the reduced state and the irreducible whole.

Statement 5: Decision Processes are substantive approaches based on defining states and predictive options towards a new state. Within chaos exists patterns that can only be defined over time, states exist within these patterns only for the time that they can be observed, blink and a new state emerges. An approach must be conducive to states that have no discernible relation with another state except that they exist within the same chaos.

These five statements are cause to reflect on the philosophical implications found in complex situations. From these statements the following propositions are provided:

- Ontological implication, decisions are conceptually bounded – within both a rational, empirical structure that can create both a paradoxical and dichotomous conditions. Consider a decision that has well defined procedures generated from a symmetrical problem set, established from years of experience, being applied in an asymmetrical situation (the one-off scenario)
- Epistemological implication, decisions are non-monotonic – decisions should focus on not how knowledge is formed but how understanding is generated. Consider corporate knowledge developed from years of experience coupled with the intuitive nature of time spent at the job for extended periods.
- Teleological implication, decisions must be longitudinally sound (across time) – goals are not projected rather they are obtained. Consider projecting a design out 5 years and making decisions within the design exclusive of the environment it finds itself in, eventually over time the environment is going to impose change onto the design that will either cause the design to fail or build something that no longer meets the need of the environment.
- Methodological implications, decisions are intuitive – Beyond the empirical and rational probabilistic methods factors should be consider that provide a catalyst for pragmatic decision making. A general methodology must be adaptive to changes in both the spatial and temporal dimension of the environment and not imposed a dominate process (one-size-fits-all) on the situation.

From these implications a general theme, decisions are pragmatic can be inferred and from this theme a thesis stated:

> Execution of a decision process in a complex situation that sustains a continuous selection of plausible possibilities will avoid the need for probabilistic states as a catalyst for the decision maker.

9.4 What Is Known

Currently there are two interpretations for decision theory, a third hybrid pragmatic/practical interpretation, a recent addition, represents a maturity in the former interpretations garnered over time.[4] Until recently conventional wisdom had decision theory interpreted two ways: normative (also referred to as prescriptive) and descriptive. These interpretations provide theory for not only how a decision should be made (normative) but also how a decision is actually made (descriptive). Normative interpretations are empirically based and lend themselves to quantified techniques. Specifics of this method include detailed experiments for data gathering and bounding to assure accuracy of the decision. Descriptive interpretations (while no less quantifiable) lends itself to qualitative methods, where interviews and surveys are typical data gathering techniques and relying on rational bounding methods for affecting a decision. Both methods place the decision maker outside the decision space for an effective decision, arguably allowing for an unbiased perspective of the problem. The third hybrid interpretation, however, places the observer inside the decision space and process focused relying on participant expertise as the dominant method for data gathering and a bounding construct. This method supports a near-real time component to the decision process, where bias and change is obviated through expertise, and by default highlighting the influence of time on a decision. To understand how this may be relevant to the decision process it is necessary to review the maturation of decision theory over the time.

The most notable and first philosopher for decision theory was Condorcet who hypothesized three stages in the decision process: discussion, clarification, and choice, Condorcet, ([1793] 1847). Condorcet posited the intransitive nature of decision making, a very enlightened position, suggesting decisions were temporally bounded. Dewey veered off from Condorcet's theory and postulated five consecutive stages: a felt difficulty, the definition of the character of that difficulty, suggestion of possible solutions, evaluation of the suggestion, and further observation and experiment, John Dewey's ([1910] 1978), this segue from Condorcet's, potentially process primed approach, integrated well with the dominant influence that sciences was having on the discipline, the demand for quantifiable knowledge as a basis for decisions and dominated research in the decision sciences to-date. Simon modified Dewey's work to facilitate decision within organizations and proposed three phases: intelligence, design and choice, Herbert Simon (1960). Brim further added to Dewey's work to: Identification of the problem, obtain information produce possible solutions, evaluate solutions, and select strategy, Brim et al. (1962). Minzberg introduced a radically different perspective to the decision process migrating from the more sequential process to a simultaneous process. Minzberg suggested that the

[4] Naturalistic Decision Making (NDM) emerged in the early 1990s in a response to the lack of decision making capabilities found in traditional methods for: "contextual factors that affect the way real-world decision making" Zsambok (1997). NDM liberated researchers from relying on the analytical and prescriptive methods of traditional interpretations.

decision process was a parallel process rather than a sequential process and parsed his phases aligned with Simon's work: identification, development, and selection, Mintzberg et al. (1976). The distinction for Minzberg being that each phase is a routine and that their progression through the process is circular rather than linear.

The next paradigmatic change occurred in the late 1980s with a more contextually influenced approach. Naturalistic Decision Making (NDM) advanced the decision approach with their self proclaimed paradigmatic shift: making the observer the center of the decision space (Klein et al. 1993; Lipshitz et al. 2001; Hoffman 1995; Lipshitz 2001). Time pressure, uncertainty, ill-defined goals and high personal stakes were the contextual factors influencing decisions in crisis (Klein et al. 1993), and the expert are key for effective decision making. The idea that context must be established from within the domain rather than an empirically or rational extrapolation was a radical departure for the decision sciences, however, the method was exclusively for time sensitive decisions, in other words a crisis.

The common denominator for each of these interpretations is a phenomenological based construct absent Condorcet's original contribution, the dimension of time. Klein and group had made a major step towards the role of including the observer into domain; however, it appeared the paradigm had not shifted sufficiently to break from the stronghold of a substantively primed interpretation into a process primed interpretation for which it showed promise.

Multi-Disciplines deal with complex situations…
…And complex situations are constantly changing.

A foundational concept of multidiscipline is *gestalt*; it implies that wholeness can never be recognized within the disparate disciplines that attempt to study the whole – they are too embedded within their own traditions. Instead emphasis is placed on the concept of '*meta*' (OED 2008), or universal as suggested by Plato[5] (Gutenburg 2009); more comprehensive concepts that convey the idea of wholeness to its parts; either concept holds true for decision making.

The myriad fields of study that have coalesced around the engineering management; fields such as: system of systems, systems engineering, modeling and simulation, and program management to name a few; all have the common goal of interaction of people and technology, and as Kovacic et al. (2006) suggest are complex situations.

A complex situation is when dynamics, emergence and uncertainty exert sufficient influence that absolute statements, "the probability of making erroneous knowledge claims" Sousa-Poza (2006) or, indefeasible conditions eliminate the ability for singular decision Complex situations postulate that all dimensions of a situation are the domain of the decision space and need to be addressed together (ref Fig. 9.1) Knowledge, although certainly not the only source of invariability, is a critical contributor in complex situations and plays a key role in understanding, or

[5] Plato describes a universal as common qualities that exist among particular things. In decision making within extreme uncertainty a universal would describe commonalities amongst all interpretations of theory to derive at a decision [8].

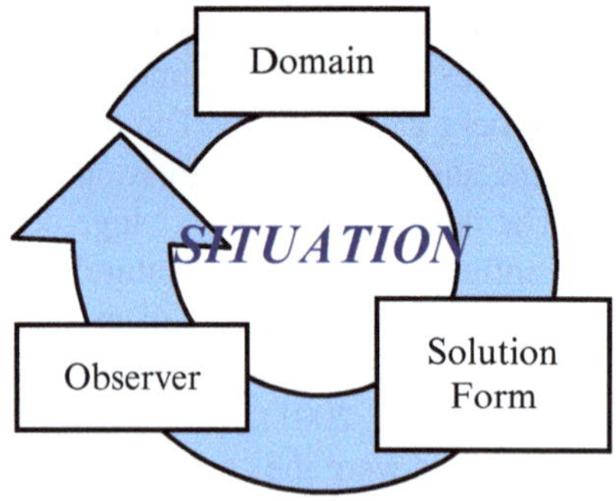

Fig. 9.1 Complex situation

rather the lack of not-knowledge plays a key role in not understanding. Padilla (2007) indicates that understanding is a function of both knowledge and not-knowledge (a lynchpin for decision making); not-knowledge can accurately be described as the unknown/unknowns and where intuition/and instinct would exert influence in understanding. Since it is understanding that must be achieved for a decision in complex situations, complex situations would necessitate a new construct for substantiating knowledge that accommodates both intuition and instinct and provides for methods for better decision making.

The history of decision making, with the exception of Condorcet, was predicated on the concept of knowledge and that knowledge is either known or knowable. The methods stipulated a decision was generated after the careful deliberation of the facts at hand and a quantifiable technique applied for making the best decision. It wasn't until Klein that the possibility of unknowns was introduced as a possible attribute of the decision in the form of using expert judgment in lieu of empirical or rational knowledge. Placing the expert into the decision space was a significant and important consideration because it introduces a new dimension to knowledge: the unknown/unknowns. The concept of gestalt and the inclusion of the observer/participant suggest more. The implication: that more than knowledge is necessary for a decision requiring a more tacit component found in understanding to affect the decision rather than knowledge alone, however, to date the interpretations support a monotone understanding. The limitation to all three interpretations is the belief that decisions are a phenomoligical occurrence reduced down to a single best solution or starting point and that as knowledge is generated understanding is increased.

By the time the third interpretation was introduced it became apparent that a foundational component was missing in the equation: the penchant for normative or prescriptive or naturalistic interpretations of decision theory in a substantive approach is absolute. However with the introduction of the third interpretation the idea that the stages of a decision process were not just sequential or parallel but based in a situation is important. The decision making stages went from a stochastic construct to a situational one defined by: situation type, suitable goals, courses of action, and relevance. This was the last piece of the puzzle to evolve the decision process under a new paradigm one that suggest that something lies beyond the known's and known unknowns but into the realm of the unknown unknowns influenced most by the participant. The catalyst for the evolution was the idea that all three interpretations are needed for an effective decision.

The paradigmatic imposition of time insists that a decision must accommodate for change, uncertainty, and dynamic emergence to truly be effective.

Para-phrased, Sun Tseng once stated: "the victorious army is the one that sets the condition for victory and pursues the condition, while the loosing army defines and pursues victory". Loosely interpreted it is more important to understand where you are going than how you get there. This is the paradigm that is adopted for integrating all three interpretations and adopting a more pragmatic approach for decision making. What Sun Tseng is referring to is a process approach, however current decision approaches rely on a substantive approach a reinterpretation of the approach highlighting the conditions for a decision rather than the approach is required. Under this modality an assessment of the stages of a decision might be re-interpreted as: *truth, plentitude, and action*; each stage having a reciprocal relationship with the other. Reciprocity encourages understanding at the same time fosters an intuitive/tacit flavor critical for maintaining the necessary temporal dimension of the individual while maintaining a coherent relationship with a-temporal aspect of the whole, key for the impetration of decision theory. Detail into each stage will provide us with the factors for architecting a decision process.

Nicholas Rescher (2000) states: "Several, if not all, of the major elements of the ontological repertoire (God, Nature as a whole, persons, material substances) are best understood in process terms...

> ...A process philosopher, then, is someone for whom temporality, activity, and change-of alteration, striving, passage, and novelty-emergence-are the cardinal factors for our understanding of the real."

Nicolas Rescher aptly highlights the need for a process approach for decision making, especially one where the decision is relegated to a person.

Evolution of these interpretations would suggest it is not one interpretation for a given scenario based on the problem but a balance of all interpretations based on the situation.

An interpretive structure must be sought to deal with the dynamic and paradoxical nature of decision. A structure that allows for:

Chaotic structures
Algedonic feedback
Non-parametric
Coherency
Intuition and instincts
Action
Conceptual analysis

The stages of RDM, are defined to account for a more dynamic and intransient conditions. *RDM is not a fourth interpretation rather a holistic interpretation of the three mainstream interpretations discussed earlier in this paper.*

At first blush when attempting to understanding RDM a natural assumption would be to infer the reverse aspect is a type of forensic approach for making decisions. Under the current paradigm this approach would make sense, linkages could be made via numerous inferences and through expert judgments a new decision made.

However, reverse literally means reversing how a decision is primed for action, start off with the decision and adjust based on the impacts of the action, this would necessitate some type of loopback to adjust for the more in-transient conditions of the complex situation and allow for a complex approach for determining possibilities, but most importantly a process approach would also imply a feedback mechanism for a continuous process. This would require a significantly robust and adaptive bounding construct more conceptual than formalized and an interpretive structure that could deal with the many possibilities generated from an initial condition.

9.5 Interpretive Framework

What do you get when you cross Rescher, Kant, Mintzberg, and Aristotle?

You get a process primed approach that accounts for the mystery of the unknown unknowns, and the benefit of continuous integration, all bounded by a collective transcendental truth: you get Reverse Decision Making

There is a systemic logic to how RDM is described in this paper, but does not imply that it is systemic. As such no transitions are provided from one component of RDM to the next to avoid the implication that this is systemic, linear, or a sequential process. Additionally, it is the purpose of the methodology to govern the enactment of RDM within a given situation and the method for this research to highlight its usefulness and the enactment to provide the necessary substantiation of the theories employed. This paper only documents the categories within the framework in the interpretive process of complex situations as well as the process itself.

9.5.1 Truth

Within the convention of mainstream decision process, identifying the problem would be the first step, in RDM it is the accepting truth of the situation. Considering the dynamic nature of the domain and varying perspective it is prudent to generate an understanding of what can be known and where the limits of that understanding exists. In addition a generalized consensus of the limits of the problem domain will provide a nascent and common point to rally around. Phenomenological understanding of the domain is critical, even more critical is dealing with the continuous process of knowing 'when' the understanding occurs. The temporal aspect of the domain demands a baseline for which change can be measured as well as where the litmus may be for triggering action.

Truth is a function of understanding which is a function of knowledge and not knowledge. Truth requires an evidentiary method for belief that is lost in methods of justifying knowledge alone. Plato's JTB provides a basis for coherent evidence yet has the fundamental flaw of not being complete Gettier (1963), the search for the condition that justifies the efficacy of JTB + in decision making can be found in Alvin Goldman's (1967) JTB + causality.

9.5.2 Causality

Goldman suggests that: "answering traditional epistemological questions requires both *a priori* philosophy and the application of scientific results. As he often puts it, Goldman's naturalism is the view that epistemology "needs help" from science". This suggests a teleological relationship to decision making and suggest the relationship is both causal while remaining reliable within the perspective of the whole. A teleological construct must be pursued that forms a reliable method for projecting into the future while maintaining a coherency to the parts that form the whole. There are a number of reliable teleological constructs to choose from that support Goldman's work, but, Aristotle's definition of causality supports Goldman's fourth condition, and provides the critical lynchpin for decision making: final cause, a teleological condition that embraces a process philosophy for obtaining goals subsuming the substantive nature to the process (Aristotle and WD Ross 1994). Final cause allows for the a-temporal existence of a bounded construct while allowing for the temporal change within its bounds. This is important when dealing with equi-probable possibilities to arrive at or near plentitude within a decision.

9.5.3 Plentitude

Mainstream decision processes use alternatives as means to quantify a decision; alternatives are down-selected based on probabilistic and judgment techniques that support a basic premise that there is ultimately one best alternative. More importantly it assumes that by choosing an alternative there is a linear effect on the outcome of the decision. This linearity eliminates, to an extent, the need for coherent transitions and minimizes the effectiveness of feedback, particularly within an algedonic loop – crucial for dynamic, changing, overwhelming conditions prevalent in non-monotonic decision space. In non-monotonic situation where each alternative become equi-probable a technique must exist that assumes that any solution is possible and equi-probable, and that the path to choice lies in the convergence on plentitude: the best of all possibilities. Imposing probabilistic methods for choosing the best possibilities is akin to fitting a square peg in a round whole – they do not fit.

9.5.4 Action

There is a participatory component to decision making, addressed in part by Naturalistic Decision Making methods, yet not sufficient for decision making in complex situations. Critical (Klein et al. 1993; Karol et al. 2006) for decision making is identifying the catalysts for action to provide the necessary metrics for judging the effectiveness of the decision by how the domain reacted to the decision.

The outcome of the decision is the definitive metric for decision making, necessitating a feedback that provides near-real time reaction to the decision. Defining the variables for the catalyst determines the baseline for deltas to be measured as well as defining the boundary of the new state created by the decision. Paithankar (2007) suggests confidence and motivation as the catalyst for a decision and the baseline for a new event. Action has a reciprocal relationship with an event, within an iterative construct set the stage for initial conditions for each event.

9.5.5 Methodology

Brewer suggests a methodology (ref to Fig. 9.1) that governs complex situations. Brewer (2008) provides four principles for governing a decision process conducive to complex situations.

- *Within a CRR, establishing comprehension a posteriori defines a justifiable basis for action a priori. (Action Principle)*
- *Within a CRR, lack of understanding justifies learning. (Learning Principle)*
- *Error between understanding and comprehensibility (complexity) justifies adaption of the CRR. (Adaptation Principle)*
- *Absent additional information, each situation is equally relevant at any given time. (Situations Principle)*

9.5.6 Process

Mainstream decision theory interpretations follow a substantive approach (ref Fig. 9.2). A tightly bounded construct from knowledge and understanding are defined simultaneously, and transitions from one substantive phenomena to the next are maintained through definable linkages.

Rescher process philosophy lends itself to a process primed approach (ref Fig. 9.3) implying that the beginning of an event comes from a process philosophy in a decision rather than a state (antithetical to complex situations), a condition necessary for most probabilistic models, and ends in an action generated from a decision. Process approach resists the need for an iterative structured identity, avoiding the static dependencies of a fixed nature and providing for achievable statements of the future. Rather a process approach assumes that a substantive iteration exists within its own definition and may not be representative of the next. Additionally a process approach makes unnecessary the need to bind both the atemporal nature of knowledge to the temporal characteristic of understanding, relegating the need for define transitions between one state and the next (Fig. 9.4).

In complex situations a process approach obviates the implication of complexity by accepting its attributes [such as]: non-ergodicity, non-monotonicity, emergence

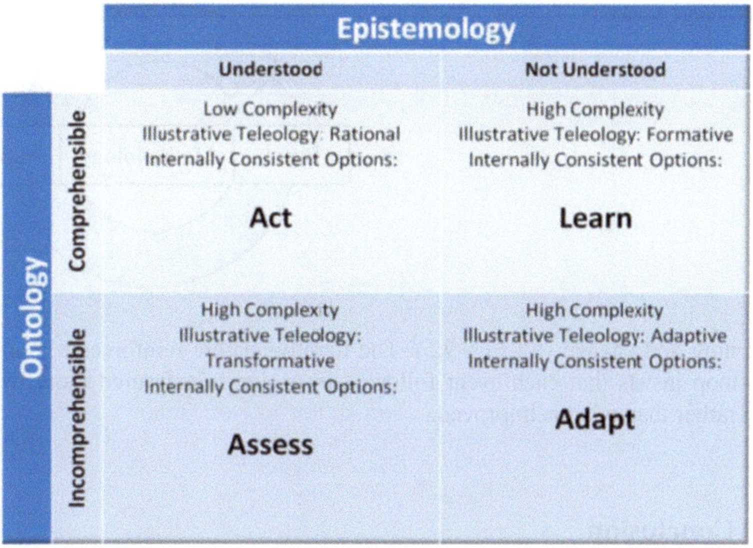

Fig. 9.2 Methodology

Fig. 9.3 Substantive approach

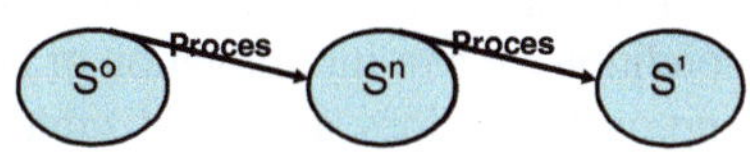

Fig. 9.4 Process approach

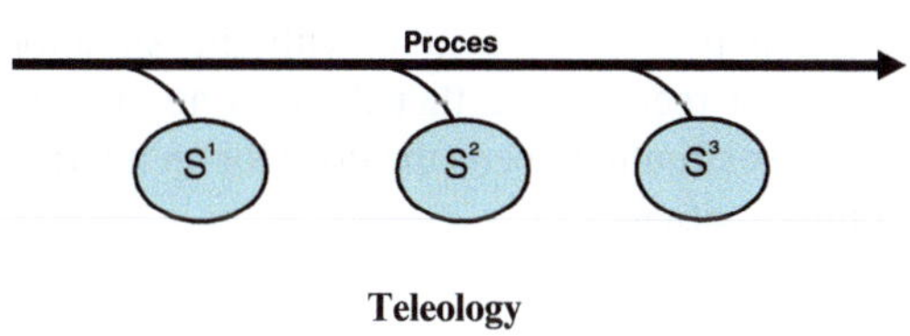

and uncertainty rather than extracting, excluding, or containing them from the decision space.

9.5.7 *Initial Conditions*

A process approach imbued with final cause relieves the need for maintaining strict correspondent traceability from one state to the next, instead implies that each event is independent of the initial conditions that created that event and action is a function

Fig. 9.5 Initial condition

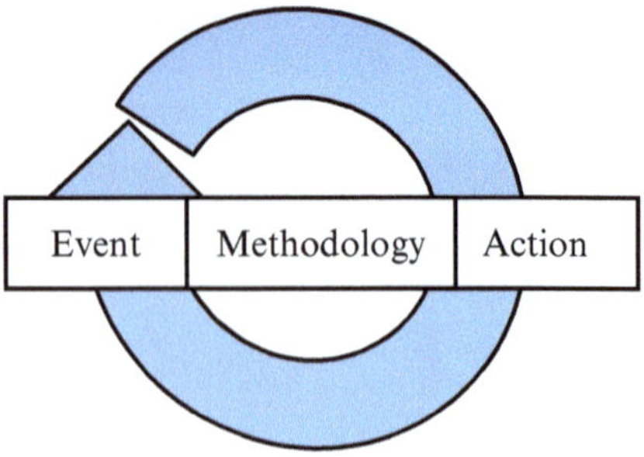

of learning and change (ref Fig. 9.5). The iterative nature reinforced via an algedonic loop insists that each event followed by action is measured more by their action rather than what is improved.

9.6 Conclusion

Given the indefeasible nature of a complex situation making a decision using mainstream decision making techniques is insufficient. A holistic interpretation of the decision space and relaxing the adherence to a substantive solution-set create the condition for a re-interpretation of how decision making is done. Adopting a process approach provides a generalizable means to assess the problem 'in situ' and to adjust accordingly, allowing for a robust selection of decision tools. Accepting that every decision beyond the initial condition is an opportunity to re-evaluate the situation as it responds to the environment provides the necessary freedom for the decision process to evolve with the situation rather than dictate the situation. Foremost recognizing that there is no single approach to making decisions in a complex situation opens the door for an interpretive framework for decisions within complex situations.

References

Aristotle Metaphysics, Book V, Part 1. Posterior analytics, Book 2, Part 11 and Posterior analytics, Book 2, Part 2. WD Ross (Translator) (1994) Original work circa 350 BCE.Retrieved 14 Aug 2010 from http://classics.mit.edu/Aristotle/metaphysics.html

Brewer V (2008) A decision making construct for complex situations. A concept paper for a dissertation topic submitted to the faculty of Old Dominion University

Brim OG et al (1962) Personality and decision processes studies in the social psychology of thinking. Stanford University Press, Stanford, 1962

de Condorcet JMAN ([1793] 1847) Plan de constitution, presenté a la convention nationale les 15 et 16 février 1793. Oeuvres 12:333–415

Dewey J ([1910] 1978) How we think. In: Middle works, vol 6, pp 177–356

Gettier E (1963) Is justified true belief knowledge? Analysis 23:121–123

Goldman A (1967) Naturalistic epistemology. Internet encyclopedia of philosophy. http://www.iep.utm.edu/nat-epis

Gutenburg Project (2009) Plato Theaetetus. Dialogues. Retrieved from project gutenburg (web site) at http://www.gutenberg.org/dirs/etext99/thtus10.txt. 18 Aug 2008
Hoffman R (1995) A review of naturalistic decision making research on the critical decision method of knowledge elicitation and the recognition priming model of decision-making with a focus on implications for military proficiency (Prepared for the University of Nottingham under sponsorship of the Defense Research Agency Farnborough and Epistemics, Ltd.)
Karol G, Shafer J, Klein G (2006) Professional judgments and "naturalistic decision making". In: Ericsson KA, Charness N, Feltovich PJ, Hoffman RR (eds) The Cambridge handbook of expertise and expert performance. Cambridge University Press, Cambridge, pp 403–419
Klein GA, Orasanu J, Calderwood R, Zsambok CE (1993) Decision making in action: models and methods. Ablex, Norwood
Kovacic S, Sousa-Poza A, Keating C (2006) Complex situations: an alternative approach for viewing a system of systems, systems, man and cybernetics. SOSE IEEE international conference. Los Angeles Ca
Lipshitz R (2001) Puzzle seeking and model-building on the fire ground: a discussion of Karl Weick's keynote address. In: Salas E, Klein G (eds) Expertise and naturalistic decision making in action: models and methods. Ablex, Norwood
Lipshitz R, Klein G, Orasanu J, Salas E (2001) Focus article: taking stock of naturalistic decision making. J Behav Decis Mak 14:331–352
Mintzberg H, Raisinghani D, Théorêt A (1976) The structure of 'Unstructured' decision processes. Admin Sci Q 21:246–275
Mitroff I (1998) SMART THINKING for crazy times, the art of solving the right problems, 1st edn. Berrett-Koehler Publishers Inc, San Francisco
Oxford English Dictionary (OED 2008) Online edition.
Padilla J, Sousa-Poza A, Tejada A, Kovacic S (2007) Towards a diagnostic framework for understanding complex situations. International conference on complex systems (ICCS), Quincy, MA
Paithankar S (2007) Radical impact of change in actions and confidence index on reverse decision making: an application based study. A thesis submitted to the Faculty of Old Dominion University
Rescher N (2000) Process philosophy: a survey of basic issues. University of Pittsburg Press, Pittsburgh
Simon H (1960) The new science of management decision. Harper, New York
Sousa-Poza (2006) ENMA 801 lecture. Old Dominion University
Sven Ove Hansson (1994) Decision theory: a brief introduction. Department of Philosophy and the History of Technology. Royal Institute of Technology, (KTH) Stockholm
Zsambok CE (1997) Naturalistic decision making: where we are now? In: Zsambok CE, Klein G (eds) Naturalistic decision making. Erlbaum, Mahwah

Chapter 10
Leadership in Complex Situations

Challenging ideas for challenging times

Diane K. Norbutus and Thomas J. Norbutus

Abstract The common understanding of position-based leadership may prevent participants from considering better ways to address the continuum of complex situations. By definition, a single person cannot know everything about a complex situation—or it would not be complex (Rittel and Webber, Policy Sci 4:155–169, 1973). This article proposes that leadership be redefined to be the act of voicing a change idea (McCrimmon, Burn! 7 leadership myths in ashes. Action Publishing, Gloucester, 2006) which means those with sufficient courage to speak up in a given situation are, in that moment, leaders. The role of executives/managers/supervisors involves not only voicing their change ideas (thus providing leadership) but also setting the conditions for more voices to be heard. The more complex the situation, the more important it is to seek all relevant perspectives. It then becomes the personal responsibility of each person to voice their change ideas. Anyone could have information, insight, or a possible approach. From this leadership in all perspective, the importance of each conversation is apparent (Stacey, Complex responsive processes in organizations: learning and knowledge creation. Routledge, New York, 2001, Complexity and organizational reality: uncertainty and the need to rethink management after the collapse of investment capitalism. Routledge/Taylor & Francis Group, London, 2010). Patterns of conversation can be noticed and named that identify an emerging way forward. Ways to embrace wide-ranging voices include utilizing the tenets of knowledge creation (Nonaka The knowledge-creating company. Harvard Business Review, November–December, pp 96–104, 1991) and open space technology (Owen Expanding our now: the story of open space technology. Koehler Publishers, Inc., San Francisco, 1997). Choosing this proposed perspective

D.K. Norbutus, Ph.D. (✉) • T.J. Norbutus
Emergent Leadership Group,105 Oxford Road, Williamsburg, VA 23185, USA
e-mail: dnorbutus@gmail.com

S.F. Kovacic and A. Sousa-Poza (eds.), *Managing and Engineering in Complex Situations*, Topics in Safety, Risk, Reliability and Quality 21, DOI 10.1007/978-94-007-5515-4_10,

opens the possibility of everyone leading, everyone contributing, and everyone more successfully navigating (perhaps repeatedly) through the complex situation.

The complexity that surrounds us is reflected in recent explorations of globalization (Croci Angelini, 2011), internationalization (Scott-Kennel and von Battenburg, 2012), investment capitalism (Stacey, 2010), information technology (Marler and Liang, 2012), engineering (Bosch-Rekveldt et al. 2011), the Occupy Wall Street movement (Davenport, 2011), Arab Spring (AlSayyad and Massoumi, 2012), and leadership to name a few. Maznevski (2011) in her article The Complexity Conundrum noted the short term focus of managers who were "simply unprepared to anticipate the impact of their decisions in a more complex world" (p. 3) and laments manager naivety in the still lingering economic crisis. Addressing leadership in this volume could help all leaders and managers deal more effectively with the complexity that seems to be increasing at every turn.

Before considering leadership in complex situations it seems prudent to first understand what is meant by complex situations. Rittel and Webber (1973) identified the difference between *tame problems*—which have fairly straight forward solutions, and *wicked problems*—which have no solution at all. They are not wicked in a moral sense, but extremely difficult to address—such as global warming, urban development, and poverty. For our purposes, complex situations would, by definition, not consist of tame problems or they would not be considered complex. Instead, complex situations would include all non-tame situations—a continuum of situations that at least initially have no clear solution. Complex situations could have a solution eventually—in effect taming them—or could be found to be wicked and unsolvable. Strategies for addressing wicked problems (Rittel, 1972) have significant implications for leaders. Three different perspectives of leadership can help us identify possible approaches across the continuum of complex situations.

This paper examines the role of leadership in complex situations starting with identification of wicked problem characteristics in Sect. 10.1 followed by ideas from the concept originators on ways to deal with them in Sect. 10.2. Some complex situations will fail to meet the extreme of wicked problems. For instance, building a skyscraper is a complex situation requiring leadership and involving a number of groups such as architects, engineers, bankers, owners, project managers, and tradesmen. However this kind of situation is solvable as evidenced by the plethora of skyscrapers in most major cities. Section 10.3 addresses non-wicked complex problems and Sect. 10.4 provides ideas from industry on ways to deal with complex situations.

In Sect. 10.5 we turn our attention to leadership. We first examine three leadership paradigms: *headship, bi-directional leadership*, and non-positional *leadership in all* for their utility in complex situations. Since it is far less understood, Sect. 10.6 grounds the idea of *leadership in all* in the theory of *complex responsive processes of relating* and the literature on *emergent leadership*. Section 10.7 identifies *knowledge creation* and *open space technology* as two ways to address complex situations. Section 10.8 summarizes leadership implications in complex situations.

10.1 Wicked Problems—The Extreme Complex Situation

A special category of planning problems have been identified by Rittel (1972) and Rittel and Webber (1973) to be wicked—that is to say: thorny and complex. These kinds of problems, societal problems, are "ill defined: and they rely upon elusive political judgment for resolution (Not 'solution.' Social problems are never solved. At best they are only re-solved—over and over again)" (Rittel and Webber, p. 160) and are called wicked, not from a moral sense but because they are particularly complex, tricky, and difficult. In an effort to clarify this situation for societal planners, Rittel and Webber identified numerous characteristics of wicked problems.

One characteristic, explained in the context of a documented reorganization, is that the understanding of a wicked problem is so closely affiliated with problem resolution that the understanding the problem depends on one's idea for how to solve it and vice versa. During the studied 2005 reorganization it was evident that every person had a different understanding of the situation and thus a different perspective on resolution (Norbutus, 2007). In this complex situation it was simply not possible to come to a common understanding of the problem which made it impossible to determine a rational way ahead using a straight forward decision model.

Since there is not a clear end point to a wicked problem resolution can *always* be better. In math there is a *right* answer. At some point however, instead of stopping because the logic of the problem determines a clear finish or end point, effort on resolving wicked problems is reduced by limits of "time, money, or patience" (Rittel and Webber, 1973, p. 162). A decision is made that the currently derived resolution is good enough in the current situation. The participants of the 2005 reorganization continued to reorganize after the main advocate left. That effort eventually reached a period of equilibrium then two more major reorganizations were attempted. There have been pauses but no stopping (Norbutus, 2007).

Wicked problems, after implementation, will result in expected and unexpected consequences over a significant period of time. People have different perspectives so when a change is viewed retrospectively, various people will note different outcomes. Because of this it is quite impossible to have an ultimate test of success that meets all the differing criteria of everyone associated. There is no ultimate test for even if one tried to test it, the conditions would have continually changed making it as un-testable then as now (Rittel and Webber, 1973). The 2005 reorganization participants wrestled with what to do, and then given the go-ahead, started to actually reorganize without testing options. Two reorganizations later, people still reflect on the 2005 effort examining its good and bad points but not considering it a solution at all (Norbutus, 2007).

Since there is no opportunity for trial-by-error outside of the problem itself there is a consequence associated with any effort to deal with the problem. An attempt to diagnose the problem is also an attempt to resolve the problem. Just as one cannot stick one's toe in the same river twice, making any attempt to resolve the problem changes the situation and further efforts will change the situation yet again with each effort counting towards improving or harming what was going on before

(Rittel and Webber, 1973). As mentioned above, there was no option testing during the reorganization (Norbutus, 2007).

Rittel and Webber (1973) posit that each wicked problem will be distinct in important ways though perhaps very similar on the surface to other situations. Because of this, planners are encouraged to delay assigning solutions to wicked problems. Though two reorganizations have occurred in the same organization since the 2005 reorganization, different people were involved and some lessons from earlier efforts were clearly not applied to latter attempts. Each effort was essentially unique.

In addition everyone experiences the resolution of a wicked problem differently (Norbutus, 2007). An academic reviewer of the 2005 reorganization remarked after reading the nine documented perspectives that it seemed each person had experienced a different event (D. Fields, personal communications, July 19, 2007).

Hard science researchers advance hypotheses that are either corroborated or refuted and the scientific community does not blame researchers if the latter eventually occurs. This, however, is not the case for planners involved with wicked problems because "the aim is not to find the truth, but to improve some characteristics of the world where people live" (Rittel and Webber, 1973, p. 167). Planners are held responsible because every attempt at resolution counts significantly which brings into the situation moral and ethical elements as expectations are created and changes made which impact people's lives. Thus wicked problems bring with them more significance than many have considered. A moral dimension is inherent in the social for as Ciulla (1998) wrote, "Ethics is primarily a communal, collective enterprise, not a solitary one. It is the study of our web of relationships with others . . . . a universe of others [is] an ethical universe" (p. 30). Re-solution of wicked problems reflect the characteristics that make these problem so deceptive and complex.

10.2 Ways to Re-solve Wicked Problems

Rittel (1972) posited that the two types of problems planners might face—tame and wicked—require two different approaches. He noted that systematic approaches to straightforward problems have resulted in spectacular results reflected in such things as the infrastructure of roads and bridges, efficient production, and scheduling. However, this often assumed paradigm based on science and engineering—does not fit the paradigm of wicked problems because "somewhere the social professions were misled ...into assuming they could be applied scientists—that they could solve [social] problems in the ways scientists can solve their sorts of problems" (Rittel and Webber, 1973, p. 160). This *serious* error has made societal problems even more difficult to address because the very nature of the problem has been unrecognized.

According to Rittel and Webber (1973) a classical systems approach, assuming the project can be organized into distinct phases, is not a appropriate scheme for a wicked problem for "one cannot understand the problem without knowing about its context; one cannot meaningfully search for information without the orientation of a solution concept; one cannot first understand, then solve" (p. 162). Instead the planning mode is a process of interpersonal iterative discourse in which problem clarity and possible solution options emerge. Some considerations suggested by Rittel (1972) to approach wicked problems include:

- Involve those who will be affected since they will have the best knowledge about the wicked problem.
- Make every part of the planning process transparent. Every step in wicked problem re-solving involves moral and ethical attitudes not scientific expertise so the problem solvers could be anyone involved, all of whom have the right to judge the re-solution.
- Allow everyone to explain to each other his/her judgment about the emerging plan. It is in the conversation that various approaches can emerge.
- Working wicked problems is always political, thus it is not reasonable to conduct stepped planning. Instead, iterative opportunities for conversation are more appropriate.
- Problem re-solving is understood as an argumentative process. This process encourages the raising of questions and issues that can be considered and with which various stands can be taken by various responsible agents in the process.

Leadership perspectives will be reviewed relative to these suggestions, but first the non-wicked complex situation is explicated below.

10.3 The Non-wicked Complex Situation

Complex situations have emerged in industries as diverse as aviation, law, medicine, finance, software development, national defense, and construction (Gawande, 2009). Technical advances and complexity in these and other professions has resulted in the emergence of highly skilled, expert individuals doing highly specialized activities either alone or with others. In many cases people, who may not even know one another, must work together doing things no one can do individually such as design and build jet airliners, navigate the legal system, provide health care, analyze the stock market, design software, prevent a terrorist attack, or build a skyscraper. This complexity has resulted in avoidable failures becoming pervasive. Gawande noted, "the volume and complexity of what we know has exceeded our individual ability to deliver its benefits correctly, safely, or reliably. Knowledge has both saved us and burdened us" (p. 13).

10.4 Ways to Solve Non-wicked Complex Situations

The traditional ways professions have dealt with complex situations has been to "encourage more experience and training" (Gawande, 2009, p. 12). We find ourselves in a time of extreme specialization where practitioners practice for years in schools, internships, and on-the-job training focusing on ever narrower content. This is a starting point. However, at some point specialization and practice are not enough.

Gawande (2009) posited that when the consequences of not getting the simple things right are significant, such as crashing an airplane or killing a patient, and there are ample opportunities for distractions, checklists can "provide a kind of cognitive net. They catch mental flaws inherent in all of us—flaws of memory and attention and thoroughness. And because they do, they raise wide, unexpected possibilities" (p. 48). For example, one of the best test pilots crashed a new highly anticipated bomber during a 1935 demonstration. The airplane was much more complex than earlier aircraft. Instead of further education and practice, other test pilots figured out the most important things to do and created checklists to ensure the basics were done. With the checklist, the plane ended up flying 1.8 million miles without one accident (p. 34). Checklists are making a difference in the medical profession as well. One Michigan hospital started in the bottom 25% of American hospitals in incidents of central line infections. After implementing a checklist program (where nurses monitored doctors and executives were directly involved in collaborative problem solving meetings) the hospital cut their infection rate to zero. Using the same process, Michigan hospital infection rates fell so low they outperformed 90% of American hospitals which avoided an estimated $175 M in expense and saved 1,500 lives in 18 months. The best applications for checklists are when the situation can be "broken down into a series of simple problems" (p. 49) such as the key steps in taking off and landing an aircraft or maintaining sterility when inserting a central line in a patient. Taking the time to make a checklist only makes sense when there are critical things to always do despite the distractions of the situation (Gawande).

Checklists help in more complex situations such as building a skyscraper. Such an undertaking requires multitude of diversely skilled people where "timing and coordination become serious concerns" (Gawande, 2009, p. 49). Construction companies have addressed the problem of complexity with checklists called construction schedules. These schedules were developed when the expertise of one *master builder* was not sufficient to deal with the complexity of the project. These schedules list the day-by-day tasks that must be done, every delivery that needs to be made, every sequence that must be followed. The schedules are developed by representatives from each of the trades involved in the project then sent to subcontractors to be double checked. This "ensures that the knowledge of hundreds, perhaps thousands [of people], is put to use in the right place at the right time in the right way" (p. 62) up front.

However, this does not address the unexpected—the challenges that had not or could not be anticipated in advance. In these cases the right experts need to work together on the issue so a communications schedule—basically a communications checklist—is used. Experts, working together can deal with the unexpected by taking into account all relevant expert views and collectively deciding a way forward. The communications checklist is the mechanism to ensure the needed coordination happens. The person responsible for providing the final answer to the problem is identified and not until the challenge is solved and everyone has signed off can the next step proceed (Gawande, 2009). A thought provoking aspect of this process is that everyone, backed up by a computer program, is expected to identify problems. In one case a worker noted that not all the I beam holes lined up and a communication task was set up to ensure not only the structural integrity issue was resolved, but any second and third order effects were considered as well. Anyone can sound the alert—anyone can make a gesture to which others would have to respond with various gestures of their own. One opportunity at a time, this type of complex situation can be solved. And the checklists have been astonishingly successful for non-wicked complex problems. Despite the fact that buildings have become increasingly complex with higher building standards, building failures are exceedingly rare—with an "annual avoidable failure rate of less than 0.00002%" (p. 71). Checklist can work in these kinds of complex situations (Gawande, 2009).

The construction schedule is a centralized, top down directive. The communications schedule is underpinned by a completely "different philosophy about power and what should happen to it when you're confronted with complex, non-routine problems" (Gawande, 2009, p. 72). It focuses attention on the micro-interactions—detailing who must communicate about what issue by when. It is directive in nature, but in this case directs that decisions are made as close to the actual problem as possible. There are implications in these Gawande's suggestions for more appropriate leadership approaches to the variety of complex situations described above.

10.5 Leadership Perspectives

10.5.1 Headship

The most commonly understood leadership paradigm is that of a person in a position of authority exerting power in an attempt to control others to solve a problem. Ellis and Fisher (1994) called this "headship" (p. 180) as opposed to leadership. Headship implies that a designated person will have all the answers that are then provided by the head and followed by followers. Ellis and Fisher noted that "while this viewpoint is a common one, it may have hindered progress in discovering the nature of leadership" (p. 181). As the complexity of a situation increases, eventually it exceeds the capacity of a single person to understand. This is reflected in the fields mentioned above such as construction and medicine. At some point this common paradigm does not cover situations that are sufficiently complex. Indeed I found in the 2005 reorganization that this perspective "does not explain or include the essential interactive nature of how [leadership] occurs and because of its systems basis, does not have a way to account for transformational/unexpected outcomes" (Norbutus, 2007, p. 173).

10.5.2 Headship in Complex Situations

Headship is by definition unidirectional—from the perceived "leader" who might more properly be called an executive or supervisor. Advocates of this perspective assume the executive has the knowledge, insight, and experience to direct the most successful and "best" solution to a problem. This view has also been termed the "heroic assumption" (Welsh, 2007) which glosses over, at best, and completely ignores, at worst, the actual complexity of the situation and various perspectives of others who will be impacted. Followers are socially compelled to take on the role of helpless recipients of the executive's decree—even if the executive has no idea what the consequences of their actions might be. Headship could work in extreme emergencies, when the executive was aware of all aspects of the situation, and when a best solution was easily apparent. However, not receiving input from others ensures the executive will only be able to make assumptions about other perspectives and skip much of the complexity actually present in a given situation. Thus, headship could be used for tame problems but is not appropriate for complex situations where there is no clear way ahead and when the contributions of impacted actors is needed for a successful outcome. In the complex situations described above, headship was not identified as part of effective re-solution.

10.5.3 Bi-directional Leadership

The bi-directional leadership perspective is more egalitarian. Ellis and Fisher (1994) wrote, "Unlike the unidirectional relationship between a head and the followers, [in this perspective] the relationship between a leader and followers is reciprocal. Consistent with the doctrine of interdependence, a leader leads at the discretion of the followers. That is, the basis for the leader's power is 'consent of the governed.' A leader may be deposed at any time" (p. 180). This perspective still has at its roots the idea that a leader is in a position of authority and has *power*. Leadership continues to be thought of as positional.

10.5.4 Bi-directional Leadership in Complex Situations

A bi-directional conception of leadership more closely follows the perspectives of practitioners actually resolving complex situations. Communication goes both up and down a hierarchy with the positional "leader" no longer expected to know all the answers. Since complex situations have exceeded one person's ability to handle, it makes sense that a leader needs to have input from others to have any hope in dealing with them. However, the challenge with this perspective is seeing the leader as the person in power. This system's based perspective assumes the leader is an outside observer/change agent able to change an organization through various points of leverage. How that is accomplished brings with it systems assumption of a monolithic organization that is, or needs to be in equilibrium. (Norbutus, 2007) which is an antiquated way to look at organizing and change (Stacey, 2000, 2001, 2003, 2007, 2010; Stacey et al. 2000) and antithetical to resolving (thus changing) a complex situation.

10.5.5 Leadership in All

The most egalitarian leadership perspective is that there is leadership potential in each individual throughout the entire group or organization (McCrimmon, 2006; Raelin, 2003; Stacey, 2000, 2001, 2003, 2007, 2010; Stacey et al., 2000). This perspective of leadership focuses attention on the everyday interactions between people in their local situations as desired by Gawande above. Leadership can emerge from anyone engaged in these interactions. The influence process of leadership creates patterns of new understanding and thus has the potential to change or transform the understanding between individuals and groups of individuals. From this perspective, leadership exists in the interactions—anyone can participate in interacting, thus anyone can provide leadership to the group (Stacey, 2000, 2001, 2003, 2007, 2010; Stacey et al., 2000). Raelin (2003) suggests that in "leaderful practice…everyone shares the experience of serving as a leader, not serially, but concurrently and collectively" (p. 5) without the constraints of power and position. McCrimmon (2006) goes even further identifying leaders as anyone with the courage to express a change idea. Thus from the leadership in all perspective there is leadership potential in everyone involved in re-solving a complex situation.

10.5.6 Leadership in All in Complex Situations

Gawande (2009) and Rittel and Webber (1973) identified leaders emerging from all levels of an organization in complex situations. In a complete reversal of paradigms, nurses challenged doctors who had not maintained sterility which helped reduce infection rates. A common laborer noted a misalignment of I beams and initiated a string of coordination efforts by engineers and other tradesmen. Rittel and Webber encouraged scientists to create a collaborative process that includes everyone who might be impacted by possible problem resolution. All these are instances of *leadership in all* being used to resolve complex situations making understanding the theoretical support for this perspective significant and relevant.

10.6 Theoretical Support for Leadership in All

Emergent leadership is focused on the behaviors of non-positional leaders—people who influence others not because of a hierarchical position but because they are sought out (Norbutus, 2012). This fits perfectly with the leadership in all characterization made above for dealing with complex situations. Emergent leaders have wisdom, knowledge of a particular subject, are accessible, and make leaderful gestures. Conceptualization of emergent leadership came from an understanding of complex responsive processes of relating theory. This theory de-centers the individual and instead focuses on the interaction between individuals. Everything follows from there—personal identity, self monitoring of behavior; personal, group, organization, and social change—all result from the conversations and patterns of conversations that emerge over time (Stacey, 2000, 2001, 2003, 2007, 2010). A key aspect of this theory is that it does not privilege leadership as being position based. Instead, leadership can be exhibited by any person through their gestures to another. The influence of an executive is greater due to perceived power differentials so an executive who gestures could receive a greater response than a person who is not in a position of authority. Emergent leadership illustrates the utility of this theoretical approach.

10.6.1 Emergent Leadership

Heifitz (1994) broke with tradition and posited a difference between the formally appointed head of the organization and a leader within the organization arguing that anyone could provide leadership within a learning community dealing with tough challenges. Kickul and Neuman (2000) argued for the existence of informal leadership networks outside the formal organizational hierarchy. Building on this idea, Dervitsiotis (2005) noted that emergent leadership is "needed to stimulate innovation through enriching and strengthening both the nodes and links of the complex web of human informal networks" (p. 925). Lichtenstein and Plowman (2009) studied the community growth of Branson, Missouri becoming a country music destination. They identifyed specific activities and behaviors "that can encourage emergence in and across organizations" (p. 617). For them leadership is not driven by positional managers but by informal leaders who come from anywhere in the organization including those in hierarchical positions (Lichtenstein and Plowman).

Norbutus (2012) conducted a heuristic study of emergent leaders informed by complex responsive processes of relating theory. The study group was asked to think about their "go to" person within their informal network who provided helpful information and advice. Study participants described characteristics of the emergent leader and roles they played which were used to formulate the Emergent Leader Questionnaire. Four general themes emerged as being most important for the emergent leader : wisdom, knowledge of a particular subject, accessibility, and making leaderful gestures. *Wisdom* was based the emergent leader being open to new information, modest, and insightful among others. *Knowledge* focused on experience, interest in the subject, as well as breadth and depth of knowledge. *Accessibility* concerned how approachable, patient, and empathetic the emergent leader was, their attentiveness, desire to help, and interest in the requester. *Leaderful acts* or gestures made by the emergent leader included responsiveness to requests for knowledge; providing valid, reliable, useful information; acting in the role of renegade from the formal hierarchy, and having shared a common degree of pain. The results of the study showed highly significant (.000) correlations of between .90 and .76 (Norbutus, 2012).

McCrimmon (2006) extended the idea of non-positional emergent leaders when he used examples of Gandhi, Martin Luther King, and Nelson Mandela to explain his distinction of leadership. All of these well-known leaders changed not only organizational but societal patterns without being in the affected organizations at all. Thus being part of a hierarchy must not be the criteria for leadership. Instead McCrimmon posits that "leadership equals having something worthwhile to say plus the courage to say it" (p. 8). The idea that leaders voice moral and ethical change ideas makes leadership a participative act in which anyone can engage. Again it is found that leadership is in all.

Clearly leadership scholars are beginning to consider the importance of leaders outside a positional paradigm. This is consistent with the theory of complex responsive processes of relating that actually explains the leadership process.

10.6.2 Complex Responsive Processes of Relating Theory

Stacey (2001) articulated the theory of complex responsive processes of relating informed by complexity science making it well suited for the examination of leadership in complex situations. The theory abandons the idea of autonomous individuals, instead acknowledging that "individuals and group form and are formed by each other simultaneously.... The fundamental motivator of human behavior is the urge to relate" (Stacey, 2003, p. 411). The process of gesturing (possibly a leadership act) and responding (possibly another leadership act) are the micro-exchanges that can result in new patterns of understanding and action. The notion of self organization is not about individuals self-organizing, it is about emerging themes that organize experience. Leaders often identify and name these themes. From this perspective there is not an outside observer who can change an organization. Instead organizational members both observe and participate in organization together. Anyone actively participating in the conversation of the organization is a potential leader. An executive can make a gesture that, if responded to in certain ways, would be considered a leaderful act. Because of their position, this gesture might result in a significant response (emerging pattern) from organizational members—but it also might be ignored. Every organizational member can choose to respond to leaderful gestures no matter its origin. They can make gestures of their own which can either sustain known patterns of organization or contribute to change (Stacey, 2003).

The power of the theory for leaders is in focusing attention in an effort to work toward common understanding. Leaders can reinforce a common understanding of established patterns or make gestures toward creating new patterns. Considering the quality of one's conversational life becomes a very important leaderful activity. From conversation can emerge surprising new patterns (creativity, novelty, and innovation) however, from this perspective one cannot know in advance what patterns will develop and who will provide the leadership at any given time. Griffin (2005) described this writing, "Leadership themes emerge in the ongoing process of group interaction in which personal and collective identities are iterated and potentially transformed" (p. 29). Stacey (2003) identified activities appropriate for a highly uncertain world when he wrote that quality actions create "a position from which further actions are possible....[and] are moral and ethical in nature"

(p. 420). Consistent with the characterization of wicked problems in which "the planner has no right to be wrong" (Rittel and Webber, 1973, p. 166), Stacey also noted, "in an unpredictable world, the outcomes of an action cannot be known in advance. It is necessary to act and then deal with the consequences....one is not absolved of responsibility simply because one does not know the outcome" (p. 420). From this theory that acknowledges the unpredictability over the long term of social interaction requires an adjustment of just what control means. With this theory control is understood as a form of socially mediated constraint not an absolute. This is unnerving for an anxious population but a much more realistic way of viewing the unfolding future and consistent with Taleb's (2007) black swan effect of improbable events. The implications for leaders (at all levels) upsets the normal interpretation of positional control as an unexamined assumption of compliance. Stacey summarized,

> the main implication of the complex responsive processes perspective is the way in which it refocuses attention, not on what members of an organization should be doing, but on what they are already, and always have been, doing. If there is a prescription, it is that of paying more attention to the quality of your own experience of relating and managing in relationship with others. This is a reflexive activity requiring each one of us to pay more attention to our own part in what is happening around us. This requires a reflective development of self-knowledge. It means taking one's own experience seriously. The reward, in my experience, is to find oneself interacting more effectively, not only for one's own good, but also for the good of those with whom one is in relationship. (Stacey, 2003, p. 422)

Thus, complex responsive processes of relating theory gives us a foundation from which to understand leadership in all. It forces us out of a comfort zone that assumes leaders can control organizations into a more useful conceptualization that leaders are participants in organizations. Leaders enable and participate in developing a common understanding knowing that every conversation is an opportunity to reinforce the status quo or co-create an innovation. Consistent with complex responsive process of relating theory, anyone can be a leader as shown in emergent leadership research and by McCrimmon's (2006) examples. The knowledge creation literature and open space technology provide practical approaches to leadership in all.

10.7 Two Pragmatic Approaches to Complex Situations

Since developing an understanding of the situation is critical to complex situations, be they wicked or just complex, it makes sense to investigate processes that focus on this capability. The knowledge creation literature provides a body of knowledge about enabling an environment conducive to creating and transferring knowledge. Essentially this is possible only through developing a common understanding. Open space technology is a process that provides structure to a fluid emergent conversation of potentially hundreds of people focused on a theme or problem. Developing a common understanding initially allows the group to determine ways ahead for seemingly impossibly complex situations. Both knowledge creation and open space technology are described below.

10.7.1 Knowledge Creation

Knowledge is defined as "the fact or condition of knowing something with familiarity gained through experience or association" (Webster). Knowledge is within a person and can be developed through social processes (association). Nonaka (1991) identified knowledge as being both *explicit and tacit.* Explicit knowledge can be put on paper, formulated in sentences, or captured in drawings. Tacit knowledge is tied to the senses, skills in bodily movement, individual perception, physical experiences, rules of thumb, and intuition. Knowledge creation is the formation of new ideas through interactions between explicit and tacit knowledge in the form of socialization (tacit to tacit), externalization (tacit to explicit), combination (explicit to explicit), and internalization (explicit to tacit). Knowledge is defined as a *"justified true belief"* (Von Krogh et al. 2000, p. 6). An individual justifies the truthfulness of his or her beliefs based on observations of the world; these observations, in turn, depend on a unique viewpoint, personal sensibility, and individual experience. Therefore, when someone creates knowledge, he or she makes sense out of a new situation by creating, committing, and justifying new beliefs. In organizations, small groups create knowledge when members share what they know as well as common values and goals in a caring, relating, and enabling atmosphere. This non-threatening, non-attribution, nurturing approach is fertile ground for ideas, thoughts, and beliefs to germinate, take root, and grow—perpetually. Knowledge creation depends on your perspective and is a craft, not a science (Von Krogh et al. 2000).

Any organization that dynamically deals with a changing environment would not only need to process information efficiently but also create information and knowledge. Leaders understand that *the one sure source* of lasting tactical, operational, and strategic advantage is knowledge especially in an uncertain environment. Instituting organizational knowledge creation would radicalize innovation and creativity. Leaders who reflectively consider the need for knowledge creation can work to eliminate both individual and organizational barriers that are ultimately de-incentivizing organizational members from performing at the highest levels. Each member of an organization has unique, personal knowledge, at least part of which is tacit and not easily explained to others. Any organization's effectiveness can be threatened by stagnation—can become rooted in processes that reinforce the status quo and do not allow time for reflective thinking on fundamental goals. This may call into question entire reason for being (Von Krogh et al. 2000).

Leaders—now recognized from the discussion above as all organizational members—can understand knowledge creation by first realizing that knowledge cannot be managed but only enabled. Although ideas are formed in the minds of individuals, interaction between individuals typically plays a critical role in developing these ideas and making them embodied experiences. For an organization based on bureaucratic foundations to evolve into an environment that can foster knowledge creation requires managers to adopt new tools, skills, and cultural values. Von Krogh et al. (2000) noted that executives can enable knowledge creation by encouraging a vision of knowledge creation—developing and diffusing a vision encompassing this new ethos; deliberately focus on the organizational conversations—which goes far beyond having a meeting to inform; rallying people or groups specially designated to encourage knowledge creation; making (by word and deed) knowledge creation an important part of the organizational experience; and perpetuating local knowledge more broadly. Von Krogh, et al. postulated that organizational knowledge creation thrives in an environment based on trust, free of personal and organizational barriers, in which organizational members take an active interest in applying the insights provided by others. This facilitates higher creative thinking; stimulates the sharing of tacit knowledge, concept creation, and justification; is essential for developing powerful prototypes; and enhances the flow of knowledge across various organizational levels. As organizational members become familiar with how and why to create knowledge they can use knowledge creation concepts to start setting the conditions for knowledge creation in their areas of influence (Von Krogh et al.) which is consistent with leadership in all.

10.7.2 Open Space Technology

Organization across the country are facing significant complexities often without situational or organizational clarity. To meet these challenges, there needs to be a way to "experience fundamental shifts of understanding, relationship, and modes of work" which is precisely what participants of Open Space Technology experience (Owen, 1997, p. 59). As people use OST to clarify their understanding of the complexities of their situation, unsurpassed levels of collaboration, creativity and commitment emerge. It sounds too good to be true, but is has proven to be the case for the past 20 years, over 60,000 times, in 120 countries—all with positive outcomes when the process is followed.

Open Space Technology is a group process, driven by passion and responsibility, which allows people facing a seemingly insurmountable task or problem to quickly become involved in meaningful dialogue. Ever evolving groups of people develop a multitude of creative solutions to complex situations in an extremely compressed time frame. The process is economical in the extreme, providing a safe place in which to learn, while appropriate structure and control emerge in the natural course of the event. The main challenges of the process are creating the space to allow self-organizing to happen and the work required before and after to leverage a one-time experience into sustainable high levels of organizational performance (Owen, 1997).

The process works for 5–1,500 people over a single day or multiple days. This process is particularly appropriate for situations with high levels of complexity, diversity, actual or potential conflict, and a decision needed yesterday. According to Harrison Owen (1997), the author of seven books on the subject, "People quiet regularly do what they know to be impossible, and having accomplished that, the rest, whatever that might be, seems much less demanding….Performance rises to the extent that people are prepared to dispense with the illusion of control and break the tyranny of the clock" (p. 31).

Open space technology is the ultimate *leadership in all* activity that clearly reveals the explanatory power of complex responsive processes of relating organizational theory (Stacey, 2000, 2001, 2003, 2007, 2010). Leaders emerge throughout the process (McCrimmon, 2006; Raelin, 2003), relating is the key activity, and the related guiding principles make the entire experience safe for everyone to create knowledge.

10.8 Consistent Approaches to Complex Situations

Leadership in complex situations is not an easy undertaking. Research shows it can be messy, confusing, infuriating, and unproductive (Norbutus, 2007). An aspect of a complex situation is that it is beyond any one person's capability to understand thus a top down headship approach is very unlikely to produce an outcome that is satisfactory to all stakeholders. Even bi-directional leadership is limited because the focal point is a positional leader who may or may not be the person who best understands the complex situation, who may or may not be best able to gesture into the situation to develop common understanding and ways ahead, who may or may not be willing to listen to others, and who may or may not be able to engender trust, honesty, personal responsibility, and a focus on people.

However, there do seem to be ways to lead in complex situations. A common element in the approaches discussed above was the idea that everyone can and perhaps has the responsibility to lead in these situations. Anyone could have information, insight, or a possible way forward. If everyone in the situation contributed appropriately to designing, planning, and implementing solutions there would be a greater chance that more contingencies would be discovered, fewer second and third order effects missed, and more buy-in created by participants.

An aspect of the *leadership in all* perspective is the importance each participant places on the quality of their conversations. Patterns of conversation are noticed and named. Everyone has the opportunity to participate in conversations that affect them—thus giving them an opportunity to justify their true beliefs. Formal and informal leaders emerge who consistently show wisdom, knowledge of a particular subject, accessibility, and make leaderful gestures.

Thus, though complex situations pose significant challenges, choosing a perspective that holds the potential to address the challenges; that takes into account the perspectives and solutions of all who have *skin in the game*; that enables novelty, creativity and innovation—is the option that has the most promise. Solving, or re-solving over and over again, complex situations is an opportunity to be inclusive, to use tools such as open space technology to get everyone involved and the insights found in the knowledge creation literature. It is an opportunity to develop more productive patterns. Choosing this kind of perspective opens the possibility of everyone leading, everyone contributing, and everyone more successfully navigating (perhaps repeatedly) through the complex situation.

References

AlSayyad N, Massoumi M (2012) Religious fundamentalisms in the city: reflections on the Arab spring. J Int Aff 65(2):31–42

Bosch-Rekveldt M, Johnkind Y, Mooi H, Bakker H, Verbraeck A (2011) Grasping project complexity in large engineering projects: the TOE (technical, organizational and environmental) framework. Int J Proj Manag 29(6):728–739

Ciulla J (1998) Ethics: the heart of leadership. Greenwood Publishing Group, London

Croci Angelini E (2011) Globalization and public administration: a complex relationship. Int J Pub Admin 34(1/2):4–12

Davenport B (2011) Occupy complexity: using complexity to examine the occupy Wall St. movement. Emerg Complex Org 13(4):87–93

Dervitsiotis K (2005) Creating conditions to nourish sustainable organizational excellence. Total Qual Manag Bus Excell 16:925–943

Ellis DG, Fisher B (1994) Small group decision making: communication and the group process. McGraw Hill, Boston

Gawande A (2009) The checklist manifesto: how to get things right. Metropolitan Books, New York

Goldstein J, Hazy J, Lichtenstein B (2010) *Complexity and the nexus of leadership: Leveraging nonlinear science to create ecologies of innovation.* New York: Palgrave/Macmillan

Griffin D (2005) Leadership and the role of conflict in processes of mutual recognition: The emergence of ethics. In: Griffin D, Stacey R (eds) Complexity and the experience of leading organizations, 1st edn. Routledge, London, pp 17–30

Heifitz R (1994) Leadership without easy answers. Harvard Univeristy, Cambridge

Lichtenstein B, Plowman D (2009) The leadership of emergence: a complex systems leadership theory of emergence at successive organizational levels. Leadersh Q 20:617–630

Marler J, Liang X (2012) Information technology change, work complexity and service jobs: a contingent perspective. New Technol Work Employ 27(2):133–146

Maznevski M (2012, March 20) The complexity conundrum. Business Week. Retrieved from http://www.businessmirror.com.ph/home/pf/8803-the-complexity-conundrum

McCrimmon M (2006) Burn! 7 leadership myths in ashes. Action Publishing, Gloucester

Nonaka I (1991) The knowledge-creating company. Harvard Business Review, November–December, pp 96–104

Norbutus D (2007) Exploring the experience of organizational transformation: contrasting episodic change with continuous change. Doctoral dissertation, Regent University, Virginia Beach, Virginia

Norbutus T (2012) Scale development for emergent leaders within an organization's knowledge node. Doctoral dissertation, Regent University, Virginia Beach, Virginia

Owen H (1997) Expanding our now: the story of open space technology. Koehler Publishers, Inc., San Francisco

Raelin J (2003) Creating leaderful organizations : How to bring out leadership in everyone. Berrett-Koehler Publishers, San Francisco, CA

Rittel H (1972) On the planning crisis: systems analysis of the 'first and second generations'. *Bedriftsokonomen* NR 8:390–396

Rittel H, Webber M (1973) Dilemmas in general theory of planning. Policy Sci 4:155–169

Scott-Kennel J, von Battenburg Z (2012) The role of knowledge and learning in the internationalisation of professional service firms. Serv Ind J 32(10):1667–1690

Stacey R (2000) Strategic management and organisational dynamics: the challenge of complexity, 3rd edn. Prentice Hall, Harlow

Stacey R (2001) Complex responsive processes in organizations: learning and knowledge creation. Routledge, New York

Stacey R (2003) Strategic management of organisational dynamics: the challenge of complexity, 4th edn. Pearson Education Limited, Harlow

Stacey R (2007) Strategic management and organisational dynamics: the challenge of complexity, 5th edn. Prentis Hall, Harlow
Stacey R (2010) Complexity and organizational reality: uncertainty and the need to rethink management after the collapse of investment capitalism. Routledge/Taylor & Francis Group, London
Stacey R, Griffin D, Shaw P (2000) Complexity and management: fad or radical challenge to systems thinking. Routledge, London
Taleb N (2007) The black swan: the impact of the highly improbable. Random House, New York
Von Krogh G, Ichijo K, Nonaka I (2000) Enabling knowledge creation: How to unlock the mystery of tacit knowledge and release the power of innovation. Oxford University Press, Oxford
Webster Dictionary Definition retrieved from http://www.merriam-webster.com/dictionary/knowledge
Welsh W (2007) Transformation changes everything: a reflexive inquiry into complex responsive processes of relating in the context of ["a very large military organization"]. Doctoral dissertation, Regent University, Virginia Beach, Virginia

Chapter 11
United States Joint Forces Command: As Wicked Problem

William O. Welsh III

Abstract This paper presents three simultaneous aspects supporting the perspective that the complexity of individual, group, organization, social aspects comprising USJFCOM in the pursuit of "aspiring-to" transformation mission responsibilities is a wicked problem. The first part presents Rittel's (Bedriftsokonomen 8:390–396, 1972) and Rittel and Webber's (Policy Sci 4:155–169, 1973) ten characteristics of wicked problems as they relate to transformation aspiration. The second part describes why individual, group, organization, social aspects comprising United States Joint Forces Command (USJFCOM) meet Rittel's (Bedriftsokonomen 8:390–396, 1972) and Rittel and Webber's (Policy Sci 4:155–169, 1973) ten characteristics of wicked problems. The final part integrates thirteen of the sixteen situated emergence themes derived, through the transformative intentionality lens of Stacey et al. (Complexity and management: fad or radical challenge to systems thinking? Routledge, New York, 2000) and Stacey's (Complex responsive processes in organizations: learning and knowledge creation. Routledge, New York, 2001) complex responsive processes of relating theory, from the standpoint of simultaneously enacted participant, practitioner, and researcher roles based on my: (1) personal participation in USJFCOM emergence 1998 to date (2007), (2) professional practitioner responsibilities as strategic development advisor and transformationalist in USJFCOM 2000 to date (2007), (3) dissertation research project of USJFCOM transformation mission responsibilities from 1999 to date (Welsh, Transformation changes everything: exploring organizational leadership challenges in an "aspiring-to" transformative environment. Doctoral dissertation, Regent University, Virginia Beach, 2007). The consequent core claim here is that coherence in understanding an "aspiring-to" transformation mission, an irreducibly wicked problem, requires consciously deprivileging the present dominant systems-based discourse, specifically

W.O. Welsh III, Ph.D. (✉)
Northrop Grumman Corp, 116 Lake View Parkway,
Suffolk, VA 23435-2697, USA
e-mail: wilycat@cox.net

S.F. Kovacic and A. Sousa-Poza (eds.), *Managing and Engineering in Complex Situations*, Topics in Safety, Risk, Reliability and Quality 21, DOI 10.1007/978-94-007-5515-4_11,

in relation to leadership, diversity, complexity, and paradox, as unsuited to the intentional pursuit of transformative outcomes.

Keywords U.S. Joint Forces Command • Transformation • Complexity • Wicked problem • Intentionality • Leadership • Coherence • Diversity • Paradox

11.1 Introduction

From 1999–2007, in conjunction with my professional responsibilities as strategic development advisor to what I now understand to be the individual, group, organization, social (IGOS) aspects comprising USJFCOM, I conducted a research project (Welsh 2007) exploring the challenges of an "aspiring-to" transformational outcomes environment. In the initial stages of this participant/practitioner/researcher experience (1999–2001) I was still a confident system-based advocate although, along with my fellow participants, I experienced a growing sense that the system-based discourse dominating the perspectives of the IGOS aspects comprising USJFCOM were proving, from the practitioner perspective, inadequate in the development of participant understanding. Although explanations for dissatisfaction with the dominant system-based discourse are complex (Welsh 2007), none-the-less, the circumstances of my experiences (specifically challenges to my professional practitioner efficacy) were sufficient to prompt a comprehensive search for alternative ways of understanding the IGOS aspects comprising USJFCOM transformation purposiveness. This quest for alternatives led almost directly to Stacey et al. (2000) and Stacey's (2001) complex responsive processes of relating theory, a theory deliberately crafted to address the challenges of transformation purposiveness.

From these participant/practitioner experiences and the researcher standpoint of complex responsive processes of relating I subsequently discovered that the IGOS aspects comprising USJFCOM intentionality manifested all of the characteristics of wicked problems presented by Rittel (1972) and Rittel and Webber (1973).

11.2 A Word About Systems

The primary literature defining wicked problems is drawn from Rittel (1972) and Rittel and Webber (1973). However, these authors appear to have never fully escaped the dominant systems-based discourse, demonstrating "theoretical developments [wicked problems] that start off with radical promise but then rapidly slip back into orthodoxy [of systems based thinking]" (Stacey 2000, p. 320). Nonetheless, to understand their characterization of wicked problems, it is important to sustain Rittel and Webber within their hermeneutic context, even as they attempt to extract their social science perspective from then prevailing attempts to model social science inquiry after that of the natural sciences. Rittel claimed that the dominant systems-based approach was problematic for developing solutions in his social science field. That this challenge still persists may substantively reinforce Rittel and Webber's contention that "Societal problems … are inherently different from the problems that scientists and perhaps some classes of engineers deal with. [They] are inherently wicked" (p. 160).

11.3 A Consideration of Wicked Problem Characteristics Linked with Insights from "Aspiring-To" Transformation Causality

In continuing to question prevailing systems-based thinking, Rittel and Webber (1973) claimed, "The professionalized cognitive and occupational styles that were refined in the first half of the century, based in Newtonian mechanistic physics, are not readily adapted to contemporary conceptions of interacting with open systems and to contemporary concerns with equity" (p. 156) They also asserted, "the problems that scientists and engineers have usually focused upon are mostly "tame" or "benign" ones … For each [such problem] the mission is clear. It is clear, in turn, whether or not the problem has been solved" (p. 160). Rittel and Webber concluded, "Wicked problems, in contrast, have neither of these clarifying traits" (p. 160).

Similarly, for human interaction as "aspiring-to" transformative causality, Stacey (2001) claimed, "Movement is toward a future … unknowable … yet recognizable … that is under perpetual construction by the movement itself. There is no mature or final state, only perpetual iteration of identity and difference, continuity and transformation at the same time" (p. 60). He continued, "Movement is for the sake of expressing continuity and transformation of individual and collective identity and difference at the same time … This is [creating] the novel, variations … never seen before" (p. 60). Stacey (2001) concluded, "Cause, [processes] of movement or construction, is … processes of micro interaction in the living present, forming and being formed by themselves, [sustaining] continuity with potential transformation at the same time. Meaning, [choice, and intention arise] at the same time" (p. 60). Thus transformation intentionality shares inextricably the same characteristics as those of wicked problems. In order to examine this proposition more thoroughly, each of Rittel and Webber's (1973) "ten distinguishing properties" of wicked problems are briefly linked to Stacey's (2001) transformation causality framework, the underpinning of complex responsive processes of relating theory.

11.3.1 There Is No Definitive Formulation of a Wicked Problem

According to Rittel and Webber (1973) "The formulation of a wicked problem *is* the problem" (p. 161). They continued, "One cannot understand the problem without knowing about its context; one cannot meaningfully search for information without knowing the orientation of a solution concept; one cannot understand, then solve" (p. 162). Rittel and Webber concluded, "The process of formulating the problem and of conceiving a solution (or re-solution) is identical, since every specification of the problem is a specification of the direction in which a treatment is considered." The intentional pursuit of transformational outcomes, knowable only after the fact, will always exhibit this characteristic.

11.3.2 Wicked Problems Have No Stopping Rule

A wicked problem is technically never "done," a solution found. "There are no criteria for sufficient understanding and … no ends to the causal chains that link interacting open systems" (Rittel and Webber 1973, p. 162). Even with the "systems" conceptual baggage in place, Rittel and Webber captured the essence of Stacey et al. (2000) and Stacey (2001) transformative causality, forming and being formed simultaneously by human interacting.

11.3.3 Solutions to Wicked Problems Are Not True or False, But Good-or-Bad

There are always multiple legitimate claims to judge "re-solution" correctness. Thus these solution claims cannot be right or wrong, but only "good" or "bad" or "good enough" (p. 162). Any transformational outcome can be simultaneously judged "good," "bad" or "good enough" depending on the perspective and power of those judging. Even these judgments are not stable over time. Additionally for Stacey et al. (2000) there is a concept of "good enough holding" (p. 391) or sufficiently reduced anxiety that is "good enough" for transformational activity to occur.

11.3.4 There Is No Immediate and No Ultimate Test of a Solution to a Wicked Problem

The test of a solution is never under control. "Any solution, after being implemented, [generates] waves of consequences over an extended—virtually an unbounded—period of time [or] yields utterly undesirable repercussions which outweigh the intended advantages" (Rittel and Webber 1973, p. 163). Transformational outcomes can only be determined after the fact. They are unknown and unknowable in advance.

11.3.5 Every Solution to a Wicked Problem Is a "One-Shot Operation"; Because There Is No Opportunity to Learn by Trial-and-Error, Every Attempt Counts Significantly

"Every implemented solution is consequential ... Whenever actions are effectively irreversible and whenever the half-lives of the consequences are long, every trial counts" (Rittel and Webber 1973, p. 163). Developing potential transformational outcomes in the face of an adversary "counts significantly" and thus is unacceptable as strategy. Mitigation with respect to wicked problems is an attempt at "good enough" approximation.

11.3.6 Wicked Problems Do Not Have an Enumerable (or Exhaustively Describable) Set of Potential Solutions, Nor Is There a Well-Described Set of Permissible Operations That May Be Incorporated into the Plan

Rittel and Webber (1973) claimed, "There are no criteria which enable one to prove that all solutions to a wicked problem have been identified and considered" (p. 164). This statement reinforces the irreducible paradox inherently present in the intentional pursuit of transformational outcomes. These outcomes cannot be known until they occur and once they occur, they cannot, by definition, represent a well-described set of permissible operations.

11.3.7 Every Wicked Problem Is Essentially Unique

Rittel and Webber (1973) stated, "By 'essentially unique' we mean that, despite long lists of similarities between a current problem and a previous one, there always might be an additional distinguishing property that is of overriding importance" (p. 164) They continued, "One can never be *certain* that the particulars of a problem do not override its commonalities with other problems already dealt with" (p. 165). Rittel and Webber concluded, "If we are right about all that, the direct transference of the physical-science and engineering thoughtways into social policy might be dysfunctional, i.e. positively harmful" (p. 165). Thus, mitigation with respect to wicked problems "is the art of not knowing too early which types of solutions to apply" (p. 164) or presuming generalization from previous experience. This, in turn, militates readily against efficiency solutions considered too early with respect to transformation intentionality.

11.3.8 Every Wicked Problem Can Be Considered to Be a Symptom of Another Problem

Rittel and Webber (1973) observed, “Problems can be described as discrepancies between the state of affairs as it is and the state as it ought to be. The process of resolving the problem starts with the search for causal explanation” (p. 165). They continued, “Removal of that cause poses another problem of which the original problem is a ‘symptom.’ In turn, it can be considered the symptom of still another, ‘higher level’ problem.” Rittel and Webber concluded, “There is nothing like a natural level of a wicked problem [and] marginal improvement does not guarantee overall improvement” (p. 165) Rittel and Webber’s explanation is fraught with rational and formative causality assertions (c.f. Stacey et al. 2000; Stacey 2001) consistent with their being under the influence of the dominant systems-based discourse. *Knowing* the current state of affairs is problematic enough in social science. The pursuit of transformational outcomes precludes knowing the “state as it ought to be.”

11.3.9 The Existence of a Discrepancy Representing a Wicked Problem Can Be Explained in Numerous Ways. The Choice of Explanation Determines the Nature of the Problem’s Resolution

Rittel and Webber (1973) asserted, “In dealing with wicked problems, the modes of reasoning used in the argument are much richer than those permissible in the scientific discourse” (p. 166). They continued, “The essential uniqueness of the problem … and [lack of] opportunity for rigorous experimentation [mean] it is not possible to put hypotheses to a crucial test.” Rittel and Webber concluded, “The choice of explanation is arbitrary in the logical sense … People chose those explanations which are most plausible … Everybody picks the explanation which fits his intentions best and … conforms to the action-prospects … available” (p. 166). “Aspiring-to” transformation requires numerous diverse considerations and understanding of the response-ability to choose. This has the consequence of expanding the potential choices of explanation that “determines the nature of the problem’s resolution” (Rittel and Webber, p. 166).

11.3.10 The Planner Has No Right to Be Wrong

According to Rittel and Webber (1973) working with or in the context of a wicked problem does not permit the same immunity the scientific community might convey to “its members for postulating hypotheses that are later refuted” (p. 167). Rather they claimed, “The aim is not to find the truth, but to improve some characteristics of the world where people live” (p. 167). In precisely the manner Rittel and Webber claimed for planners, aspiring-to transformers "are liable for the consequences of the actions they generate; the effects can matter a great deal to those people that are touched by those actions" (p. 167).

11.4 Wicked Problem Social Context and Transformation Intentionality

Rittel and Webber (1973) noted the increasing heterogeneity of Western high-scale societies as, "increasingly differentiated, comprising thousands of minority groups, *each* joined around common interests, common [cult values] and shared stylistic preferences" (p. 167) differing from other groups. Rittel and Webber elaborated, "As the sheer volume of information and knowledge increases, as technological developments further expand the range of options, and as awareness of the liberty to deviate and differentiate spreads, more variations are *possible*" (p. 167). What Rittel and Webber (1973) described as the virtue of possibility becomes, with transformation intentionality, the necessity of possibility and the recognition of diversity, difference, deviance, and misunderstanding as the basis of potential transformative causality. Intentionality is conscious choice, a uniquely human capability that establishes the basis for responsibility and accountability. Choice with respect to transformation is the purposive IGOS aspect, seeking transformational outcomes in response to dynamic environmental uncertainty. "Aspiring-to," consciously sought transformational outcomes requires IGOS aspect conduct (behavior by conscious choice) directly in the face of anxiety, threat to identity, and aversion to risk attendant to any human act into the unknown and unknowable in advance.

11.5 USJFCOM as Wicked Problem Social Context Challenges

Welsh (2007) identified sixteen situated emergent themes from the "warrior participant"[1] IGOS aspects comprising USJFCOM that illustrate a select portion of the USJFCOM-as-wicked-problem social context challenges. Due to space limitations thirteen of these situated emergent themes are addressed below.

11.5.1 Pizzas to Volkswagens (1999)

This early emergent theme comprised a metaphorical pizza factory building, prominently labeled USACOM[2] "Pizza Factory," over which a bill poster is seen applying a freshly minted USJFCOM "Volkswagen Factory" banner. Vehicles delivering various essential pizza ingredients are seen competing for space at the input end of

[1] Uniformed military members of the USJFCOM community.

[2] U.S. Atlantic Command, prior to 1999, U.S Joint Forces Command's historical predecessor.

the building. It is implied that skilled pizza making human capabilities and dedicated pizza making equipment still define the unseen building interior. Four senior executives, responsible for the new transformation mission outcomes assigned to USJFCOM are seen kneeling near the building's output end, all praying ardently, "Please Lord, make the next one a Volkswagen." This theme was interpreted as the futility of warrior participants, as organizational elites, to do the same thing in more or less precisely the same way and expecting something different to happen. The argument was understood as primarily organizational rather than specifically operational. At least one inherent weakness in the storyline of this theme has been recognized. Pizza making, while clearly distinct in many ways from Volkswagen making, is nonetheless a production process. Subsequent individual/group warrior participants have argued convincingly that, except in the broadest sense, a combatant command headquarters is not a producer of war-fighter capability; it is a user (c.f., Worley 2006). With respect to the theme, the case could be made that USACOM was never a unified pizza factory. After 1998, however, the USACOM/USJFCOM headquarters staff members were the entirely otherwise-experienced overseers of numerous producer cottage industries: former chairman of the joint chiefs of staff-controlled activities such as the Joint Warfighting Center, the Joint Warfare Analysis Center, and the Joint Communications Support Element.

11.5.2 Ritual Decapitation (2000)

It is a salient feature of military organizations to routinely rotate uniformed workforce members, typically on a 24- to 36-month cycle. While all uniformed members can be rotated, the effect of this activity is most keenly felt within the organization at the mid-manager level (commissioned and noncommissioned) up through the elites to the ultimate organizational decision maker, the USJFCOM Commander—thus the term *ritual decapitation* for the senior leaders are the *head* of the organization. Whether changing command at the highest levels or merely changing out within the majority of the organization, this process is considered routine for the military, headquarters included. The expectation within all such military organizations is that the incoming member, by virtue of standardized structures, processes, and procedures together with recognized processes to amend the same, will quickly orient themselves (with some assistance from sponsors) to the particular challenges of their new situation. Thus, officers assuming command of an Army brigade, Air Force wing, or Navy group know and understand, within well established parameters, the breadth, width, and depth of the responsibility they will be assuming. What cannot be known in advance is the interactional effect of the new person with the new group and vice versa.

In a traditional military organization, this ritual is, in many instances, intentionally disruptive, allowing the new leader to assert his or her insights with respect to the task at hand, the unit's ongoing mission. However, USJFCOM warrior participants stated that there were many creative ways for a unit to sustain a particular proven way of conducting itself, sometimes in the face of a new leader's intentions to change. This is possible because each person is situated in a web of interrelationships and each decapitation event can potentially serve to sustain the interrelationship web and/or potentially transform some or the entire web of interrelationships. At the time this theme was recognized, the conversations did not reveal the paradoxical potential with respect to the webs of interrelationships. These events appeared as discrete unto themselves, as the removal of a part from a whole. Nonetheless, just recognizing the situated nature of the event did lead to a better appreciation over time of its consequences.

11.5.3 The Collegiality Factor (2001)

Two different general types of power conduct were apparent within the uniformed member organizational citizen group, which can be described as within service and between services. While there were infrequent exceptions, in general, uniformed mid-manager-level or above organizational decision makers of the same service would, amongst themselves, frequently but appropriately disagree publicly with each others' positions on work-related subjects. However, in public conversations between members of different services, warrior participants not in the conversation claimed they experienced an almost overly polite deference to others' opinions, especially when the conversants were of the same power position or rank. This is not to imply that different positions were not taken, only that the differences were rarely confronted. This was most especially true at the most senior levels where a special coded language of disagreement could be observed whereby the most senior organizational decision makers would actually praise most effusively the ideas they tended to disagree with the most.

The consequence of this cross-service coding was that the public perception of these exchanges was an imperturbable collegiality and the disposition of the real issues of contention remained ambiguous to those experiencing the conversation looking for clues to these positions. Warrior participants of that time speculated that the purpose of these types of relations may be driven by each manager's lack of familiarity with the details of every other service's reward and performance evaluation process and the absence of joint reward and performance evaluation processes. General agreement (though varied in degree) among the warrior participants emerged on this point. The conduct experienced indicated the default position in these between-service member relations was to first do no harm; most frequently, this resulted in taking no action at all to speak about, much less resolve, important individual/group/organizational/social issues.

11.5.4 The Heroic Assumption (2001)

The heroic assumption theme actually originated with a USJFCOM senior mentor, an extremely accomplished PhD and retired Army General, who was frequently overheard in this time/space period to observe that the typical senior military officer and supporting staff arriving for joint force training "make the heroic assumption that they already know what they are doing, most often before they can even know what they are expected to do." In USJFCOM warrior participant groups, this concept was quickly extended to describe anyone arriving at USJFCOM who believed that they had any idea what the command's mission actually required.

For an extraordinarily rich variety of reasons, humans have attempted to dominate other humans since before recorded history. The critical aspect with respect to the "aspiring-to" transformative environment is in Burrell's (1988) notion that "knowledge … is enmeshed in a power field [and that] the petty malices of those who seek to dominate mean that knowledge itself is part of the domination" (p. 228). This assertion gives support to the further notion that knowledge *is* power and that knowledge is to be controlled if power is to be gathered and preserved. This claim is clearly antithetical to the development of transformational outcomes, thus *heroic*. In the face of the aspiration for creativity, novelty, and innovation, power elites will persist in their hierarchically reinforced domination prerogatives built, as Burrell noted, into the very group/organization/social processes.

Thus, those making (or enabling) the heroic assumption in the "aspiring-to" transformative environment induce at least two additional challenges into that environment: (a) significant resources will be engaged, attempting to control what cannot be controlled and (b) these attempts at controlling the uncontrollable promote a futile and unnecessary resistance that further wastes resources and opportunities for potential transformation.[3]

11.5.5 Higher-Level Processes and New Assembly Area (2001)

Although a government Senior Executive Service civilian initiated the conversation regarding transformation production processes, several groups of warrior participants within the disparate communities (including especially the Joint Force Trainer community) and USJFCOM Headquarters staff immediately contributed to the conversation and, in many respects, quickly influenced it. These conversations resulted first in the definition in four categories or aspects of higher level processes: broadly define U.S. Department of Defense capability requirements typically culminating in documents such as the National Security Strategy, National Defense Strategy, and National Military Strategy; combatant commands' current and future joint and beyond joint capability development requirements for which USJFCOM must produce products and services; inter- and intra-community processes that directly produce products and services to meet combatant commands' current and future joint and beyond joint capability development requirements; and finally, warrior participant identified processes that facilitate and enable (e.g., personnel administration, infrastructure and logistics, security, and communications/information technology).

This theme, that there are higher level processes that inform USJFCOM activities, survives explicitly only in the command historical documents. Implicitly, however, the theme has survived in every succeeding regime's rediscovery of the need to

[3] Hypothetically, extreme forms of exercised superordinate power may subsequently manifest in equally extreme forms of resistance with the potential for transformational outcome as revolution.

address the fact that a combatant command (US Atlantic Command) was simply redesignated USJFCOM without a discussion of what that meant (though it has never yet been formulated that way). Most regrettably, these conversations to date though continuing focus on processes have all been systems-based and thus about control, privileging the largely unexamined notion that combatant command headquarters staff processes, structures, and conduct are also naturally and precisely the right ones for transformation production integration.

11.5.6 *"But, this is a military organization!" (2001)*

This theme emerged in three aspects: (a) in response to the uncertainty engendered in confronting the transformation mission; (b) after September 11, 2001, in reasserting the essential primacy of "militariness" to meet the uncertainties of the Global War on Terror ("designated a *war*, after all"); and (c) in the subsequent ongoing operations in Afghanistan and Iraq. By their use of this expression, it became evident that warrior participants were reinforcing to those around them an affirmation of what is, to them, obvious. New warrior participants (generally still within the first 45 days of their arrival in the community) could imagine no other characterization. Warrior participants with longer experience in the community appeared to be less positive and generally exhibited more willingness to debate alternative individual/group/organization/social interactional possibilities.

11.5.7 *"It's just a basic leadership issue" (2002)*

This theme emerged from the same three aspects as "But, this is a military organization!" detailed above. From the warrior participant conversations of the time (2002), this theme appeared on the one hand to be a final hopeful assertion of militariness framework, within which the warrior participants could reasonably claim they knew what they were supposed to do. On the other hand, this theme also appeared to represent an apparent personal resignation to the uncertainties and frustrations commensurate with the theme's assertion. By this theme, in perfect adherence to hierarchical tenets, warrior participants transferred accountability for outcomes they could not control themselves to the senior warrior participant elites. The theme persists, potentially as both the last vestige of warrior participant hopefulness in their assertion of militariness and a possible lament at the emerging perception of their assertion's inherent limitations. There appeared to be some relationship between this theme and prior claims by certain command-experienced participants captured in the heroic assumption (see above) and "But, this is a military organization!". These three themes appeared to be part of a framework of social/psychological accommodation. This framework appeared to mutually reinforce the tenets of these three themes that support, by their inherent militariness, warrior participants' sense of self/identity and thereby reducing their anxiety in acting in the USJFCOM "aspiring-to" transformative environment.

11.5.8 *Transformation Changes Everything (2002)*

This theme emerged from warrior participant conversation as a result of my introduction, as strategic development practitioner, (a) the tenets of Stacey et al. (2000) and Stacey's (2001) complex responsive processes of relating theory and (b) the Office of the Secretary of Defense and Joint Staff Training Transformation (T^2) initiative within which USJFCOM (specifically the Joint Warfighting Center) was directed to develop and manage deployment of a new Joint infrastructure for training transformation. Promoting understanding of complex processes of human relating provided useful and necessary distinctions between rational and formative causality perspectives and a transformative causality perspective appropriate to at least the potential for success in an "aspiring-to" transformative environment. The advent of the T^2 initiative repopularized, at least with respect to the Joint Force Trainer, part of USJFCOM transformation mission responsibilities. By their conversational struggles, the warrior participants manifested that they were quite conflicted about the simultaneous convergence of these ideas leading them to confront the notion that transformation changes everything.

Although simple to state, the theme is inherently vast in scope and extremely complex in detail. The words, borderline tautological, claim coherency with the understanding of transformation as the generally irreversible conversion of nature, function, and/or condition. Even conceptualizing such a notion requires risk taking, threatens identity, and induces anxiety. Nonetheless, stated this way, what could be known and obvious is made conscious and emphasized. This theme continues to serve as an effective barometer of warrior participation in the "aspiring-to" transformative environment. Those confounded by the words retreat from contemplation of their strategic significance. Those rare few energized by the words evangelize in quixotic fashion the profound truths contained therein. The jaded many, either never having engaged (still confounded) or worn out simply from keeping up with the dominant systems-based perspective are seemingly content to simulate the tactics of World War I land battle and dutifully obey the rules of trench warfare, producing like results until they can intellectually retire (metaphorically dead) from the field.

11.5.9 *The Kandahar Road (2003)*

Patton (2002) quoted apocryphal Grand Canyon hiking advice. He wrote, "Don't mistake a clear view for a short distance" (p. 39). In terms of this theme's emergence, this quote could also be rendered as, "Don't mistake the desire for a clear view for a short distance." The dominant systems-based perspective, with its cybernetics-motivated focus on control, produced an interesting effect within warrior participant conversations. Having identified for themselves (or having outside consultants identify for them) the normative ideals they perceived were related to their current challenges, warrior participant group conversations, encouraged by warrior participant elites, fed in varying degrees by their personal heroic assumptions, routinely equated the identification of necessary action with its accomplishment. The premise appeared to be fed, in turn, by traditional hierarchical hegemonic expectations of sovereignty and obedience in the application of appropriate controls for that believed to be in need of control and controllable.

Fielder and Reynolds (1988) political film, *The Beast of War* presented the story of a ruthless Soviet tank commander, the Beast, who has just savagely murdered a village elder under his tank treads. After this operation, as their unit is in the process of returning to Kandahar, the commander, through his own arrogance, directs his tank driver to take the wrong fork in the road. The tank commander and crew, now lost in a valley with but one exit, are set upon by revenge-seeking mujahideen warriors from the village in which the elder was killed. During the run of this valley gauntlet, the tank commander turns on members of his crew, killing one in a paranoid rage and leaving another booby-trapped in an unsuccessful effort to kill the pursuing mujahideen.

The theme metaphor derives from the next scene. The tank, containing the commander and remaining two crew members, at last emerges from the valley and proceeds at full speed across the open desert. Soon, to their great and visible relief, they spot their elusive goal—the road to Kandahar. For a few moments, the tank's elated occupants cheer, the horror of the past days fading in the contemplation of safety. Suddenly, the commander, perched atop the turret, orders, "Stop tank! Stop tank!" The camera pans up and over the commander's view to reveal they are poised on the edge of an impassable abyss nearly rivaling the Grand Canyon in depth and breadth standing between their precarious position and their goal, the road to Kandahar. They now realize they will have to go back, through their pursuers in the valley gauntlet, if they are to survive.

The theme (understandably but conveniently) dispenses with the political aspects of the war, the situational insanity of the tank commander, his murderous actions, and their consequences (except as these reveal survival of a heroic trial). Instead, the theme, as it emerged in warrior participant conversations, concentrated on the emergence of clear purpose following great struggle, the enthusiastic embracing of self-evident solutions, and the frustration and bewilderment that results from the realization that the challenge is impossibly complex and can only be approached by reengaging in risk and the corresponding anxiety and threats to identity that this entails. This theme gave great expression to warrior participant issues with their responsibilities to USJFCOM's transformation mission and the development of an appropriate "aspiring-to" transformative environment.

11.5.10 The Hegemony and the Anarchists (2003)

This theme emerged from two ultimately converging conversations. The first conversations were warrior participant extensions of the heroic assumption, "But, this is a military organization!" and "It's just a basic leadership issue" themes. The warrior participants reflected their natural expectation that USJFCOM was, in fact and after all, a military organization, a hierarchy with power/authority emanating from the top and distributed in predictable ways down through the hierarchy. They frequently expressed puzzlement resulting from their experience of the warrior participant elites (senior command decision makers) commanding certain actions to be taken, whereupon these directives, the warrior participants then witnessed, were either ignored, deferred in perpetuity, or otherwise subverted.

The second set of conversations that ultimately resulted in the emergence of the subject theme originated in my strategic development practitioner attempts to develop an understanding of this warrior participant experienced phenomenon and subsequent exposure to a rich array of international relations theories, which deal specifically with issues of power, balance of power, sovereignty, authority, and anarchy.

Understanding the theme's emergence did require understanding of hegemony (characterized by sovereignty and the expectation of obedience, the quite natural state of a military hierarchy) and anarchy (characterized predominantly by anarchic or self-help behaviors). The invocation of international relations theory perspectives brought an important distinction regarding the conceptualization of anarchy. The term was initially most commonly understood as a state of society without government or law. However, another perspective entirely complementary to an "aspiring-to" transformative environment through the perspective of complex responsive processes of relating, maintains that anarchy is a theory which regards the absence of all direct or coercive government as a political ideal and that proposes the cooperative and voluntary association of individuals and groups as the principal mode of organized society ("Anarchy" n.d.). Extrapolated into the warrior participant conversations, this latter definition gave meaning to the uncomfortable tension between (a) the warrior participants' natural affinity for hegemonic hierarchical authorities and attendant presumptions of stability in the maintenance of the status quo and (b) an emerging understanding of IGOS aspects appropriate to an "aspiring-to" transformative environment.

11.5.11 Being There, Going There (2003)

This theme emerged from within USJFCOM warrior participant conversations in late 2003/early 2004 in conjunction with a USJFCOM command-wide horizontal integration initiative—the Command Objectives Review. This theme appeared to be based loosely on the road to Kandahar as warrior participants' struggled to accommodate senior executive heroic assumption-based expectations. Simply stated, command senior executives wanted the results of being there (the ability to provide the commander with the assumed-to-be positive results of various staff information and so-called knowledge management processes as these might relate to the command's so-called production processes) without going there (to whit, actually developing the perspectives, processes, and procedures capable of promoting transformative outcomes in an "aspiring-to" transformative environment). Under pressures to be there, warrior participants rapidly developed satisficing responses designed to convince command senior executives that their expectations were being met. The theme is a simple juxtaposition of states understood to be mutually exclusive. It captures warrior participants' frustration with unrealistic senior elite expectations that once the need for a process, procedure, or outcome was identified, immediately or sooner the results of the process, benefits of the procedure, or availability to the outcome were expected at hand.

11.5.12 Current Operations Always Trump the Future (2005)

The theme's explicit emergence in the warrior participant conversations is deeply embedded in the 1999 arguments to establish USJCOM. At that time, decisions were made and communicated unambiguously that USJFCOM, unique amongst its perceived peers, would have no operational war-fighter responsibilities (hence, no area of operational responsibility) in order to allow the command members to concentrate entirely on the development of transformed joint force capability and joint capable forces. By this action it was understood, though only rarely discussed, that this unique arrangement was necessary because emphasis on current operations always trumps the future. Some warrior participants had historically taken this stricture on USJFCOM members as a metaphorical gelding of traditional war-fighter capability development prerogatives (in effect, through traditional service-based practices, to direct perceived effective and efficient outcomes). However, the important reason for the creation of USJFCOM was that since the Goldwater-Nichols act in 1986, the emergence of joint had not occurred spontaneously from the intentions of that act; thus, USJFCOM was created to accomplish this purpose without other naturally compelling (especially to war-fighters) distractions.

The theme is another straightforward acknowledgement of the obvious to this project's warrior participants. Variations on "there isn't time for tea and philosophy when you have bad guys in the wire" abounded as the nature of the conflicts in Afghanistan and Iraq changed from the assured victory of superior forces and principles to protracted commitment to nation building and countering insurgency, civil war, and resurgent fanatical fundamentalists. The theme can be understood as a warrior participant apology for their present pragmatic inability to focus on transformation. Of course, even such an apologetic discourse masked the inherent inability of the warrior participants' underlying perspective (systems-based thinking) to actually allow them to focus on transformation.

11.5.13 Transformation as Prescribed Outcome (2006)

This theme was inextricably linked to an increasing emphasis on current operations (see Sect. 11.5.12), this theme initially manifested shortly after a commander departed in August 2005 and was reinforced with the arrival, fresh from the frustration of Middle-East war-fighting responsibilities, of a new commander. The now well-recognized consequences of ritual decapitation (see Sect. 11.5.2) were reinforced with the April 2006 USJFCOM chief of staff (COS) regime change. These two events contributed significantly to the preemptory disestablishment of the earlier COS's apparently tortured and torturing attempts to establish horizontal integration across the command headquarters membership just as these efforts were poised for at least some degree of success. The new COS adopted a now historically predictable heroic no-nonsense results-now approach, supported thoroughly by the equally predictable heroic assumptions of later COSs. An anemic transformation rhetoric barely maintained a place in the formal vision, mission, goals conversations until mid-2006 when the USJFCOM commander was challenged by his hierarchical superiors regarding the apparent absence of transformation emphasis in the command goals and objectives.

This paradoxical theme captured warrior participant frustration with transformational uncertainty, unpredictability, and consequent uncontrollability. The felt need to provide immediate tangible support to those directly engaged in the Global War on Terror and the even more immediate operations in Afghanistan, Iraq, and the Horn of Africa dominated every warrior participant conversation.

11.6 Conclusion

The irreducible individual/group/organization/ social aspects comprising USJFCOM "aspiring-to" transformational outcomes mission responsibilities meet all of Rittel and Webber's (1973) characteristics of wicked problems. The dominant systems-based discourse in social science offers no prospect of a commensurable understanding because this discourse, despite its occasional radical potential for understanding, is predicated on knowledge in advance of a definable system and an ability to control complexity. Viewed through the complex responsive processes of relating perspective, the situated emergent themes derived from certain individual/group/organization, social aspects comprising USJFCOM present an alternative potential understanding, one appropriate to the complexity of the deliberate pursuit of transformational outcomes.

References

Anarchy (n.d.) Dictionary.com unabridged (v 1.1). Retrieved 4 Jan 2007, from http://dictionary.reference.com/browse/anarchy

Burrell G (1988) Modernism, post modernism and organizational analysis 2: the contribution of Michel Foucault. Organ Stud 9(2):221–235

Fielder J (Producer), Reynolds (Director) K (1988) The beast (of war) [motion picture]. (Available from Columbia Pictures Industries/W. Washington Blvd, Culver City, CA)

Patton MQ (2002) Qualitative research and evaluation methods, 3rd edn. Sage, Thousand Oaks

Rittel H (1972) On the planning crisis: systems analysis of the 'first and second generation'. Bedriftsokonomen 8:390–396

Rittel HWJ, Webber MM (1973) Dilemmas in a general theory of planning. Policy Sci 4:155–169

Stacey RD (2000) Strategic management and organisational dynamics: the challenge of complexity, 3rd edn. Pearson Education Limited, Harlow

Stacey RD (2001) Complex responsive processes in organizations: learning and knowledge creation. Routledge, New York

Stacey R, Griffin D, Shaw P (2000) Complexity and management: fad or radical challenge to systems thinking? Routledge, New York

Welsh III WO (2007) Transformation changes everything: exploring organizational leadership challenges in an "aspiring-to" transformative environment. Doctoral dissertation, Regent University, Virginia Beach

Worley DR (2006) Shaping U.S. military forces: revolution or relevance after the cold war. Praeger, Arlington

Part III
Cases and Approaches

Chapter 12
Using Systems Design to Solve Complex Problems: Case Studies from Iraq

Nathan A. Minami

Abstract As the first decade of the new millennium demonstrated, the majority of the challenges and threats that America now faces are of a complex nature. There has been much academic thought regarding the nature of complex problems and that has produced many noteworthy attempts in both public and private sectors to develop methodologies for understanding and confronting complexity. Despite these efforts, global leaders still struggle to confront the complex challenges of the twenty-first Century: AIDS, cancer, obesity, climate change, over-population, crime, terror, poverty and an unstable and turbulent economy are just a few. This paper provides some practical ideas and methodology for approaching complex problems. While complex problems rarely have a solution, there are both good and bad approaches to solving them. Identifying optimal solution sets is critical to success.

12.1 Introduction

As the first decade of the new millennium demonstrated, the majority of the challenges and threats that America now faces are of a complex nature. Consequently, over the past 10 years there has been much academic thought regarding the nature of complex problems and there have been several noteworthy attempts in both the public and private sectors to develop methodologies for understanding and confronting complexity. Nevertheless, we still struggle to understand many of the complex challenges of the twenty-first century: AIDS, cancer, obesity, climate change, over-population, crime, terror, poverty and an unstable and turbulent economy are just a few.

Not coincidentally, the U.S. Army has also acknowledged a change in the operational environment. Specifically, the majority of the problems and missions that the

N.A. Minami (✉)
Military Science, U.S. Army, Oxford, MS, USA
e-mail: naminami@olemiss.edu

S.F. Kovacic and A. Sousa-Poza (eds.), *Managing and Engineering in Complex Situations*, Topics in Safety, Risk, Reliability and Quality 21, DOI 10.1007/978-94-007-5515-4_12,

Army is currently undertaking are so complex that identifying the nature of the problem is often elusive. Recently, Army doctrine writers proposed a new methodology to compliment the Army's long-standing Military Decision Making Process (MDMP). As indicated in the newest version of Field Manual (FM) 5.0, The Operations Process, a new methodology called "Design" identifies the importance of understanding complex problems fully before attempting to solve them using the traditional MDMP (Department of the Army 2010).

12.2 Design

While FM 5.0 provides an outstanding reference for describing the importance of design, it does not accurately describe a useful methodology for understanding ill-structured, complex problems. Further, there has not been a lot of academic discussion in military circles regarding the *how to* of design. Consequently, the following methodology was developed and used by planners in the 2nd Brigade Combat Team, 10th Mountain Division in Iraq during its deployment to Baghdad in 2009–2010. While there are several "steps" to the methodology, they were not followed rigorously in a lock-step manner, but rather were followed iteratively. The first step of the methodology was to ask some initial questions to help frame the problem for further analysis and understanding. The three primary questions that were asked were: What is the current state of the environment? What is our desired end state? What are the tensions that are preventing us from transforming the system from its current state to the desired end state? While these three questions might seem straightforward and simplistic, when dealing with complex problems they usually are not. Consequently, they merit extensive energy and time to answer. In-depth discussions through working groups and consultation of subject matter experts were needed in all cases to answer these questions. In some cases, a month or longer of research and reflection was needed to answer these questions.

Once initial responses to these three questions were formulated, a variety of systems thinking techniques were utilized to help understand the problem, or specific aspects of the problem, in more depth. Techniques that the design team used included critical thinking using the *Elements of Thought* where a list of critical questions are used to help better understand the problem's context and most salient characteristics (Elder and Paul 2006). In some cases System Dynamics was used as a means for understanding the system's structure. This was specifically useful for complex systems where dynamic feedback loops were present that made traditional linear thinking methods less useful (Sterman 2000). Another technique that was used was Systems Architecture which focuses on understanding the concept that maps the function of a system to its form within the larger context and interfaces of the system (Maier and Rechtin 2002). A fourth technique that was used was the Systems Decision Process which consists of many subordinate methods such as stakeholder analysis, functional analysis and value modeling (Parnell et al 2008). Finally, as the first case study will show, Network Analysis was also used as a methodology to help understand how key actors associate within the context of a larger problem (Brandes and Erlebach 2005).

Finally, the output of the design process was an articulate problem statement as well as a solid definition of both the current system as it could be observed as well as a description of the systems' desired end state. Equally important, the output included a detailed articulation of the difference between the two systems and what obstacles stood in the path of system transformation. Finally, the final output of the design process included a definition of the boundaries and limits of the system in regards to time, space, structures, people and institutions, as well as identification and description of all factors that could affect the system. The remainder of the paper will discuss how the design process described above was used by the 2nd Brigade Combat Team, 10th Mountain Division in Iraq to help better understand two highly complex and mission critical problems and develop practical solutions to the problems.

12.3 Case Study 1: Building Unity in Iraq

As the Brigade Combat Team examined the challenges in Eastern Baghdad in early 2010, one of the most complex and least understood problems was the myriad of organizations and other players that affected the security, economic and political environment. By early 2010, the level of violence had receded to manageable levels and the Iraqi Security Forces generally had control of the security structure for the area. Nevertheless, the perception was that despite improved security conditions, there were still a multitude of schisms within Iraqi society with many different actors that appeared to be openly or passively hostile towards each other. This prevented progress. Consequently, the Brigade put together a working group to address this problem.

The team began by attempting to answer the three critical questions: What is the current state of the environment? What is our desired end state? What are the tensions that are preventing us from transforming the system from its current state to the desired end state? The answer to the first question was that there was little evidence that Iraqis from different religious sects, tribes, political parties, and various security, government and entrepreneurial organizations were working together to achieve any kind of social, economic or political progress in Eastern Baghdad. The desired end state was clearly for Iraqis from these various groups to work together to implement practical solutions to problems in lieu of resorting to violence or ignoring problems all together. But the tensions and barriers that were preventing transformation from the current state to the desired end state were less obvious.

In efforts to identify these barriers, the team began to identify the various divides that prevented the Iraqis from working together. The one that was most often cited and recognizable was the general divide between Sunni and Shi'a groups. But other divides such as disenfranchised Sunni, Shi'a and Kurdish sub-populations existed because many of these sub-groups do not have equal political access due to a lack

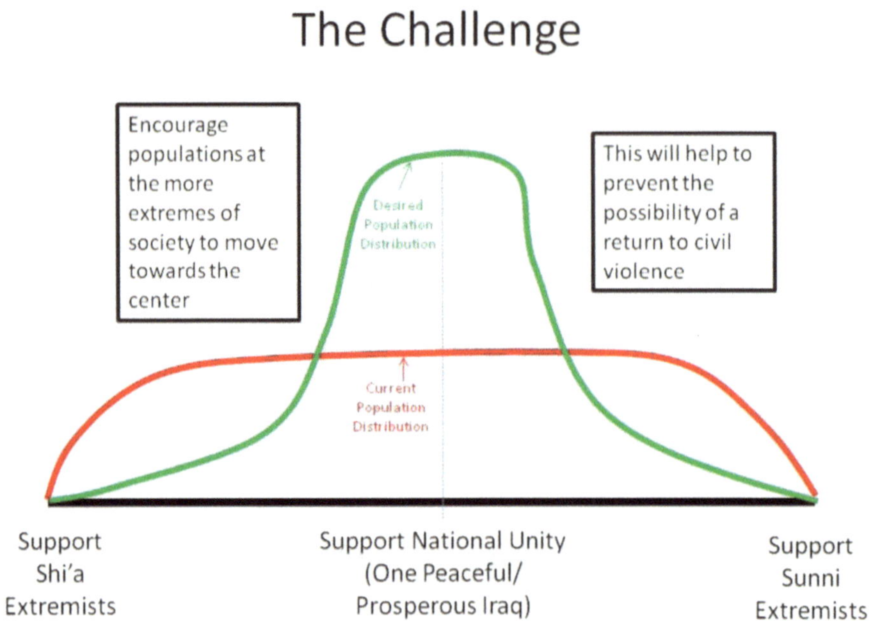

Fig. 12.1 The challenge

of balanced government representation. Other schisms included lacking minority rights, unequal wealth and benefits distribution, un-reconciled past grievances between various tribes, political, governmental, and ethnic groups, and external influences such as Al Qaeda and Iran that encourage movement towards political extremism. While all of these divides served as barriers to political, economic and social progress, they were exacerbated by myriad entrepreneurs who take advantage of these schisms for their own benefit.

Because of the many diverse schisms in Iraqi society, the team eventually acknowledged that it would be difficult, if not impossible, for any one effort to address even the most important divides. Consequently, the problem was reframed. Instead of looking at how to address the many schisms individually, it was determined that focusing on the problem in a positive manner would be more helpful. Therefore, the simple framework shown in Fig. 12.1 was developed. Instead of viewing the problem as a diverse collection of many fragments, it was reframed to look at the population on a diverse scale from those supporting extremism on various ends of the spectrum to those in the middle who support national unity and a peaceful and prosperous Iraq. While detailed statistics were not collected, it was assumed that the population followed a distribution that was generally uniform in nature. The problem therefore, was framed as developing a movement that would adjust the distribution from one that was generally uniform to one that followed more of a bell curve. Following the bell curve, the number of people who support Shi'a or Sunni extremists would be reduced and the population supporting national unity would be increased.

Once this initial framework was developed, subject matter experts and stakeholders ranging from professors in the United States to local Iraqi leaders were consulted to identify their perceptions of the problem. One theme that resonated strongly from these sources was that local Iraqi leaders did not view the problem as one of reconciliation; they believed that this goal had already been achieved. But, as they pointed out, reconciliation does not imply harmonious cooperation or even consultation on the biggest issues. They proposed instead that what was really needed was *itihad*, the Iraqi word for unity.

Again, the initial problem was adjusted to account for this key stakeholder perception. Extensive research and theoretical development regarding the *itihad* concept was then conducted which included various lessons and concepts from the general reconciliation, conflict management and negotiation, and peace and stability operations literature. While this was all helpful, the preponderance of the concept was derived from Malcolm Gladwell's book on Systems Thinking, *The tipping point: How little things can make a big difference* (Gladwell 2002).

Gladwell's ideas were critical in helping to better understand the problem and eventually develop a solution. Gladwell describes the tipping point as the moment in time when a thought, product, or virus instantly takes hold and exponentially spreads (Gladwell 2002, 12–13). Using this idea, the team determined that the initial idea for this project was to determine how the idea of unity among the people of Eastern Baghdad could become like a virus and spread throughout the region at an epidemic rate. In addition, two key concepts adopted from Gladwell's work were contagiousness, the understanding that even the smallest things can have big consequences, and that positive changes do not happen incrementally, but rather in one critical moment in time (Gladwell 2002, 9–13).

Other concepts provided by Gladwell were also important. Gladwell describes three types of people who are critical for creating an epidemic (Gladwell 2002). The first of these is a *maven*. According to Gladwell, mavens gather knowledge about a particular subject, product, or matter. Mavens are obsessed with a particular topic. The second type of individual is a *connector*, who has an astonishingly large number of personal acquaintances. Connectors can rapidly pass anything to dozens, if not hundreds, of people quickly due to their vast number of contacts on any given day. The third type of individual is a *salesman*. Salesmen are unusually persuasive and can easily sway people in one direction or another (Gladwell 2002, 30–88).

Other important concepts in creating a tipping point are *context* and *stickiness* (Gladwell 2002, 89–192). Context refers to how and which small aspects in the environment play the largest role in determining our identity, how we behave, and whether or not epidemics occur. Stickiness refers to the ability to parcel information in a manner that it is so appealing and enticing that people readily accept it.

Finally, an important concept in Gladwell's work is that epidemics have a surprising paradox. This paradox states that in many cases it is necessary to create several tiny movements before one large movement or epidemic can take hold. This is considered a paradox because while a concept or idea generally explodes exponentially in one single moment in time, it generally takes many smaller efforts for the larger desired outcome to take effect. This implies that while the "tipping point" takes affect rapidly, the period of preparation prior to this explosion may be longer than desired.

Using these tipping point concepts as a starting point, the BCT staff developed a phased methodology. This methodology allowed the team to frame the problem in a manner that could be easily understood. The first step was to identify as many mavens as possible who could help the team to better understand the problem. The second step was to understand and develop the precise message that the BCT needed to infuse into the population. The next step was to develop the context and determine which aspects of the environment weighed most heavily on the problem. The fourth step was to determine what would make the message stick, or rather how to get Iraqi leaders to adopt the message as their own. The final step was to determine which connectors and salesmen among the two million inhabitants of Eastern Baghdad would be most important for spreading the unity theme.

Based on the theoretical construct for this project, a campaign plan was developed over a series of meetings that included representatives with diverse competencies from the Human Terrain Team, Information Operations, Psychological Operations, Provincial Reconstruction Team, the Intelligence Section, the Operations Section, and representatives from all maneuver battalions. In addition, maven input included consultation with professors and PhD students, subject matter experts, a review of existing reconciliation and stability operations literature, and local Iraqi leaders from the community who were believed to have in depth knowledge regarding the problem. The most important aspect of maven input was what the group received from local leaders. Despite the wide range of expertise on the team between operators with multiple tours in Iraq, the HTT and the PRT as well as input from academically knowledgeable individuals outside the BCT; ultimately it was the local leaders who best understood the intricacies of the myriad challenges, disputes and concerns of people in the region.

Two central objectives were developed as part of the design process. The first was to encourage Iraqis to support national unity and a peaceful, prosperous and unified Iraq within Eastern Baghdad. The second objective was to encourage the population to reject extremist positions, work to resolve conflicts and disputes peacefully, and to build trust between various entities in the community. In addition, three phases were developed for the campaign. The first phase consisted of preparation, or development of the campaign plan itself. This included holding weekly unity working groups with key staff members to develop the theme, context and stickiness as well as to gather maven input and to identify critical salesmen and connectors in the community. The second phase was to employ Gladwell's (2002) paradox of the epidemic which consisted of creation of small movements that would allow the team to inject the theme into various segments of the population. The concept for this phase was to initially invite Iraqi mavens to a series of meetings where the theme, context and stickiness of the message could be refined. At some point, local salesmen and connectors would be invited to the meetings so that the theme could be injected into these key personnel and subsequently spread among the population. During these initial meetings, the plan called for development of constructive measure for diverse leaders from different communities to work together so that trust could be created. There were also two important decision points during this phase of the campaign. The first was to determine which Iraqi leaders to empower during the meetings so that those with good ideas would not be stifled by others with more *wasta* (standing or respect among peers). The second decision point was to determine which Iraqi leaders to empower with responsibility for continuing and leading the *itihad* process.

Following the campaign plan concept, the unity epidemic would be ignited during the third phase. During this period, Iraqis would take full ownership of the movement and U.S. leaders would attend meetings only to monitor and offer suggestions or assistance when asked. In some cases, it was conceived that this might include providing funding for unity projects. The end state for the *itihad* movement was that the citizens of Eastern Baghdad would overcome existing tensions and disputes and support a peaceful and unified region within Iraq. In addition, the end state envisioned citizens working together to resolve conflicts peacefully, with trust and mutual respect existing between all groups and among the majority of the population.

Once the problem, purpose and initial campaign plan construct were developed, the working group began to focus on developing the theme and addressing context and stickiness. With regards to theme, based on the input of various mavens the central theme or message was focused to place a more positive emphasis on creating trust and unity. In doing this, the ideas of faith and hope were also introduced as being central to making trust and unity possible. Some of the ideas that followed from this theme were that working together will create a prosperous future for families, children and future generations. Another sub-component of unity is that communication between various groups is important and that although there will be many disagreements, mutual respect is a critical component of creating and maintaining civil accord. Also supporting unity is transmitting the message that many are stronger than just a few, and that by harnessing the collective capabilities of the whole economies of scale can be harnessed to increase the prosperity of all. This implies that if the group is to stand united, they will be successful, but if not they are likely to fail to create the positive social and economic conditions desired by the majority.

Once this theme was developed and refined, the group examined the context of the problem in order to understand what small conditions in the environment might serve to assist or hinder efforts to promote the message of unity in Eastern Baghdad. In doing this, the group focused on identifying the environmental factors that serve to separate or prevent people from working together, and which factors unite the population. Some of the factors that were identified as driving wedges among the population are massive problems with garbage littering the streets and excessive amounts of graffiti in many neighborhoods. These factors were considered to be similar to what Gladwell (2002) described in the case of New York City crime where unbridled amounts of graffiti in the subways spurred lawlessness and criminal activity (Gladwell 2002, 5–8). Other factors that were determined to contribute to the problem were a general disconnect between the government of Iraq and the majority of the population and a lack of trust that cause people and groups to focus on their own welfare and not the collective good of the whole community. Some final contextual factors that serve to divide the population are a lack of people knowing their neighbors who live in adjacent communities, a factor that creates ignorance and serves as a barrier to communication and unity. Complementing this is a state of widespread corruption and a lack of governmental follow through which prevent groups from building trust and working together.

The working group determined that stickiness was the most difficult aspect to address. Some of the ideas that the group developed that might help to make the theme stick were that working together will enhance the quality of life for all and that cooperation is the path to developing economic prosperity. Other ideas that could help to make the message stick were demonstrating to the population that they all have a common enemy, namely extremists and Iranian influence that seek to usurp their hard earned freedoms. Other ideas that the group developed that would help to make the message of unity stick were harnessing traditional Arab values of honor, pride, heritage and self worth, as well as developing mechanisms (small victories) to show the population that working together leads to success, and that success breeds success.

The final step in concept development was to determine which of the many influential leaders in the community were essential to invite to the unity meetings. The goal was to keep meeting size to approximately 20 people to allow for all invited to be able to participate actively in the meetings without extending the length of the meetings to something unmanageable. In addition, it was decided that keeping the group relatively small would enhance the ability of group members to develop strong relationships at the meetings. Consequently, a list of the most influential leaders in the operating environment was collated based on nominations from various staff members, subordinate units, and local leaders themselves. This list of influential leaders included each leader's primary contacts with other leaders in the community. This allowed for the network diagram shown in Fig. 12.2 to be developed. In addition, special emphasis was placed on choosing leaders who had already demonstrated they are willing to place the greater good in front of their own personal interests.

While it was clear that all of the leaders nominated were important and influential (most individuals we determined to be all three: mavens, connectors and salesmen), what was less clear was how they were all inter-related. The network diagram produced by the BCT's Human Terrain Team showed that once all of the various leaders and their contacts were mapped out, that community leaders were naturally aligned along four general groups: tribal leaders, religious leaders, Iraqi Security Force leaders, and governmental/political leaders. The software package (originally designed for mapping criminal and insurgent networks) used an internal algorithm to determine how the various leaders were inter-connected and which were most important for connecting all of the disparate leaders together in a unified network. These dynamic connectors/salesmen included leaders from all four general groups and were determined to be of great importance not because of the large number of associations within their own communities (or groups), but because they were key for linking multiple diverse segments of the community together (linking different

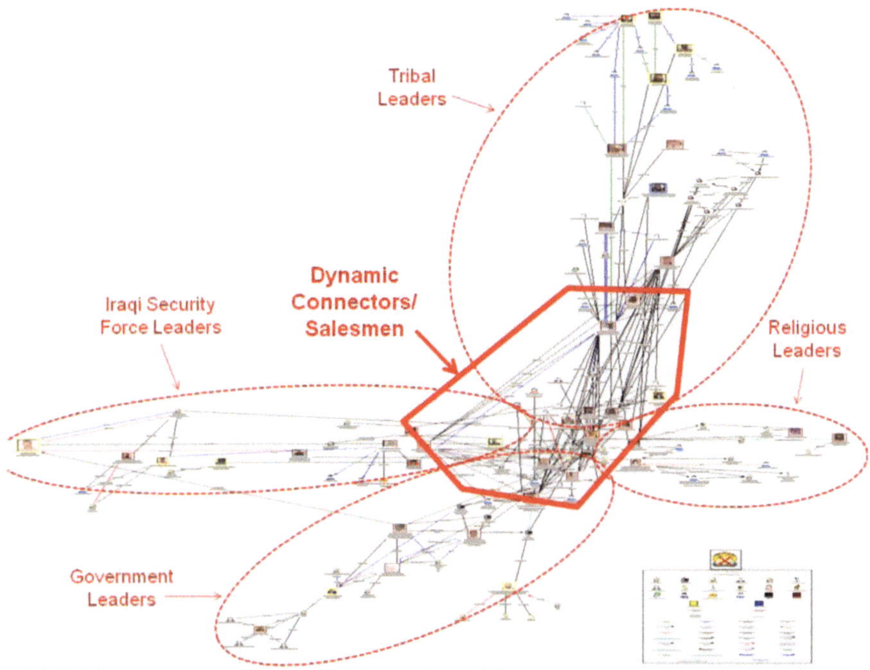

Fig. 12.2 Network analysis

groups together). While there were some exceptions made for particularly influential leaders, the majority of the people invited to the meeting came from the dynamic connectors/salesmen portion of the network diagram shown in Fig. 12.2.

While the initial concept was to invite only a small number of local mavens to initial phase two meetings, delay of the Iraqi national elections and other pending changes in the operational environment forced the initial concept to be changed. Consequently, approximately 20 individuals were invited to the first meeting who were believed to be so important that many embodied all 3 characteristics: that of a maven, connector and salesmen. In addition, the initial goal became more ambitious: to infuse the message of unity to the group at the first meeting. The first meeting began with introductions of the BCT Commander and Division Deputy Commander by the BCT's PRT representative who knew and had already built a relationship with most of the invitees. This was followed by opening remarks by both leaders, and then the BCT commander welcoming guests and inviting everyone to introduce themselves. This was an important part of the initial meeting as the meeting included such a diverse group of leaders from all segments of society that all did not know each other prior to the meeting. Following introductions by all invitees, the BCT commander delivered the unity theme. Once the theme was delivered, the BCT commander invited the guests to share their suggestions and comments. After a lengthy discussion the group broke for lunch, after which all were solicited for ideas and suggestions for the next meeting.

The central theme of the message delivered was that we (the U.S. Army) have asked you all to come together because you are all noble, influential leaders who are critical to building prosperity in the region. In order to create prosperity, you must communicate w/ each other, create mutual respect, work together, and stand united against external forces that want to create unrest. Vignettes used to describe this message were the success of recent efforts to develop democracy, freedom, elections, the economy and essential services. The message was delivered in a positive manner where special effort was made to ensure that the theme did not come across as belittling or talking down to the guests. Rather, it was delivered in a collegial manner that focused on all of the positive progress made to that point and showed how this progress could be expanded by increasing efforts to work together. In addition, it was emphasized that we viewed everyone at the meeting as equals, and stressed this point by ensuring there was no table or head in the seating arrangement. Instead, sofas were used and were placed in a square so that no invitee would feel more or less important than another.

Other important points that were made by attendees at the meeting were that improvements in essential services for people in their communities are not visible to the average citizen; this includes water, electrical and sewage services. In addition, political reforms to represent and work for all the people are slow to happen and the majority of the population is skeptical about the new government's abilities to help its citizens. The group emphasized that political reform will lead to an inclusive, secular government working for the betterment of all Iraqi's. Other concerns that were raised by attendees included a lack of Government of Iraq interaction with local communities to address issues, the local leaders and the people; especially in rural areas. Many attendees expressed concerns with economic development; that there is a need for improvement with businesses, jobs, salaries and standard of living. Similarly, corruption throughout all levels of government institutions was another concern. Specifically, its negative impact on prosperity and the ability to positively change the social system, which includes the delivery of essential services.

One of the most critical outcomes of the meeting was the group's combined understanding that they have the collective power to solve many problems themselves. Specifically, by coming together they can produce a non-partisan unified voice to the Government of Iraq (GoI) that represents the views of 2.5 million inhabitants. In addition, the group realized that their own power and resolved several problems at the meeting. Two examples were getting the government to conduct more rural visits to assist with citizen needs (fixed by a GoI rep at the meeting) and gaining amnesty for certain individuals in rural tribal areas (fixed by an Iraqi Division Commander at the meeting). Through this dialogue the *itihad* meeting group began to slowly develop trust among each other, a trust they realized would help them to withstand various pressures and external forces that will certainly be presented in the future.

Another issue raised by attendees concerned removing political hate and sectarian violence from the Friday prayer messages at Shia and Sunni mosques and

Picture 12.1 Unity conference attendees

churches. The group determined that this would be an important part of building unity in the region. A portion of the conversation by ISF and GoI officials focused on the positive cooperation with tribal leaders in helping within the areas of security, political unrest and assisting with economic development. The group recognized that the tribal leaders play an important part in carrying the message of unity to a majority of Iraqis and they needed to be incorporated into the way ahead.

The reaction of the invitees far exceeded the expectations of the BCT working group. While many invitees took the opportunity to express their concerns in the area, nearly all agreed exuberantly that the concept of unity and the *itihad* forum was one that needed to continue. Several members of the group offered to host the next meeting, and the group began working together to develop ideas for future agendas. Indeed, they insisted that a focused agenda be developed and agreed that hard solutions must result from future meetings. In addition, the group also identified and recommended including other influential members to help with issues in subsequent meetings. Consequently, while the BCT's working group had envisioned it taking several meetings for the unity idea to take hold, instead, it was immediately accepted by the group in a manner that showed the theme was much more contagious than expected. As of August 2010, the Unity Concept had spread well beyond the Brigade Combat Team's sphere of influence and was being implemented by Iraqis, across Baghdad. As of this writing, the verdict is still out on whether or not the movement will help to reconcile the deep divides inherent in Iraqi national politics (Picture 12.1).

12.4 Case Study 2: Re-entering the Triangle of Death

Following Iraqi National Elections in March 2010, the security situation in Iraq was relatively stable. There was, however, a need to gather more information in the area south of Baghdad. Consequently, Task Force (TF) 2-14 IN from the 2nd Brigade Combat Team, 10th Mountain Division withdrew from eastern Baghdad and moved to Victory Base Complex to assume the reserve mission for United States Division-Center. The battalion was then subsequently attached to one-third Advise and Assist Brigade and assigned the mission of developing information and atmospherics in the area south of Baghdad that was previously known as the "Triangle of Death."

Not only was this problem immensely complex, but there was also little time to conduct design to help better understand the problem. It was clear that the most challenging aspect of the current state of the system was that little information was available on what sort of enemy activities, if any, were occurring within the area immediately south of Baghdad. Part of the end state was also clear, the Task Force needed to be able to provide higher commanders with a clear picture of what sort of events were occurring, and if the population was supporting the Government of Iraq or other nefarious organizations.

As the Task Force's plans group began to conduct the design methodology for this problem, several salient challenges became apparent. First, the area that TF 2-14 IN would be operating in was within another battalion's area of operation, and while 2-14 IN would not work for another battalion, some relationship had to be established to prevent fratricide and ensure unity of effort. Second, the Task Force had to build a working relationship with the U.S. advisor teams that were already assigned to advise and assist the Iraqi Army Brigades that TF 2-14 IN would be partnered with in Southern Baghdad. Third, the battalion had to ensure that it quickly grasped the reporting requirements and standard operating procedures of the new brigade to which it was attached. Fourth, the Battalion had to determine ways to engage the population that previously had been overwhelmingly hostile towards U.S. forces and were believed to have felt a sort of victory that U.S. forces had finally departed their area. Fifth, the battalion would have to operate out of Baghdad and drive for approximately 1 h on potentially dangerous roads where roadside bombs were infamous for killing American Soldiers. Sixth, and potentially most difficult, the Task Force would have to overcome potential perceptions of abandonment and re-engage the Iraqi Army and Police units operating in the area after their former partners had already left them.

Further complicating this final point was that by 2010, all U.S. operations in Iraq with few exceptions had to be partnered with the Iraqi Security Forces. In this case, TF 2-14 IN would conduct partnered operations with the 23rd and 25th Iraqi Army Brigades of the 17th Division. Typically, it can take several months to develop a good relationship with Iraqi counterparts to the point where mutual trust is formed between the two groups, but in this case a relationship would have to be established in just a few days. These relationships would also have to be built on multiple levels, between the TF 2-14 IN battalion commander and his Iraqi brigade commander

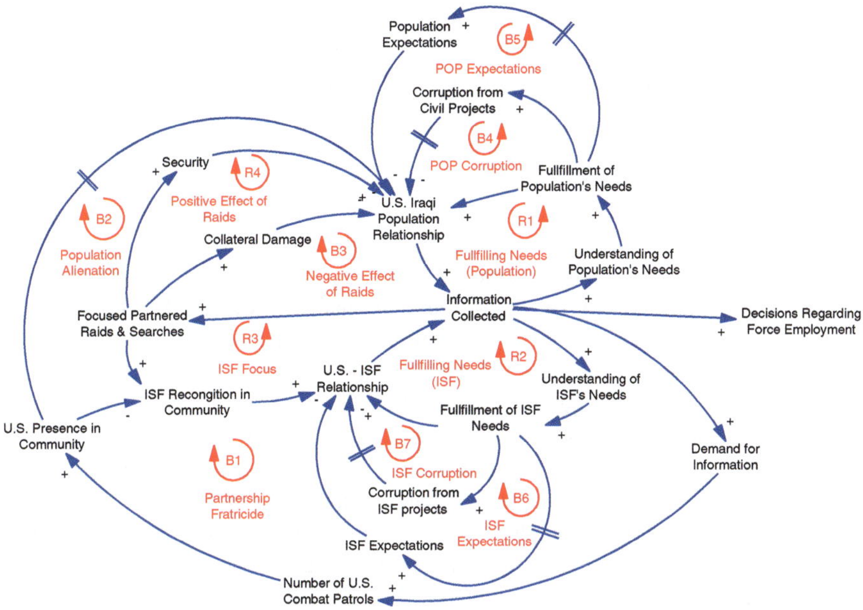

Fig. 12.3 System dynamics model

counterparts, as well as between U.S. company commanders and Iraqi battalion commanders and U.S. platoon leaders and various Iraqi patrol leaders.

A System Dynamics model (Fig. 12.3) was created to help better understand the problem. While mathematically based System Dynamics models can be created and used to drive simulations which often enhance decision making, in this case time constraints and many uncertainties dictated that a qualitative model would be most beneficial. Specifically, the model in Fig. 12.3 is useful because it provides a visual tool that increases understanding of the dynamic feedback between variables and the effect of delays in the system.

The team identified four critical reinforcing loops. Reinforcing Loop 1, or R1 (Fulfilling Needs- Population), describes how as more information is collected, U.S. forces gain an improved understanding of the population's needs and therefore, are able to meet these needs. This leads to an improved relationship with the population, and ultimately, more information collected. R2 (Fulfilling Needs-ISF) shows how the same process works with regards to relationship building with Iraqi Security Forces. R3 (ISF Focus) describes how the collection of more information leads to increased effectiveness of partnered raids and searches, which leads to increased ISF recognition in the community, an improved relationship between U.S. forces and the ISF, and ultimately more information. R4 (Positive Effect of Raids) builds on the same concept as R3, except that it recognizes that as partnered raids and searches is enhanced, security improves, which also improves the relationship between U.S. forces and the population, which leads to increased information collection.

While the model clarified the importance of focusing on these four reinforcing loops, development of the model also showed that there were several balancing loops that needed to be considered. Balancing Loop 1, or B1 (Partnership Fratricide), describes how as more information is collected, the demand for information increases, which increases the number of U.S. combat patrols needed to collect information. This leads to an increased U.S. presence in the local community and a decrease in ISF recognition in the community which leads to a weakening U.S. relationship with the ISF and less information collected. B2 (Population Alienation) builds on the same general structure as B1, except that it recognizes that as U.S. presence in the community increases, the relationship between U.S. forces and the population deteriorates due to the perception of U.S. occupation, which decreases the amount of information collected. B3 (Negative Effect of Raids) describes how as more information is collected, enhanced raids and searches with Iraqi partners could be conducted of suspected residences. This however, leads to an increase in collateral damage (sometimes things as simple as broken doors or simply unwanted presence in a person's house count as collateral damage) which erodes the relationship between U.S. forces and the Iraqi population, and leads to less information collected. B4 (Population Corruption) and B5 (Population Expectations) depict how as more information is collected which allows the fulfillment of the population's needs, over time this can lead to a decrease in the relationship between U.S. forces and the Iraqi population which is eroded due to the corruption that occurs from civil projects and increased expectations of the population. These loops are further complicated by the delays that occur between when the population's needs are met and increasing expectations, as well as between the corruption that occurs with civil projects and the resulting decrease in relationships between U.S. forces and the Iraqi population. Finally, B6 (ISF expectations) and B7 (ISF corruption) are similar to B4 and B5, except that they recognize the negative impact of the ISF's involvement in corruption and their increasing expectations that comes with U.S. fulfillment of ISF needs.

Based on this model a number of guiding principles were developed for this operation. First, it was noted that taking advantage of the four balancing loops would be vital to mission success. This meant focusing on building strong relationships with the population and the ISF. Based on information collected conducting focused, partnered raids and searches of suspected enemy locations, and putting the ISF in the lead on all partnered operations, making them the clear face of any success. The principle for avoiding the negative implications of loops B1 and B2 was simply to minimize the number of combat patrols conducted and ensure that each combat patrol had a clear mission, purpose and objectives so that the positive effects of the patrol would outweigh any unattended negative consequences. The principle developed for countering the potentially negative effect of B3 was to only conduct raids and searches when very good information existed, and then to always but the ISF in the lead during the operation. Coaching the ISF to be respectful of the residences they were searching would also be necessary to avoid potentially negative consequences. Loops B4 through B7 were the most challenging loops to address. Like any country in the Near East, corruption is rampant in Iraq, and decades of government handouts under the Hussein regime created an atmosphere where the population constantly expects increasing services from the government. To counter the negative effects of these loops, it was noted that the short duration of the mission (a few weeks) would help to prevent these factors from negatively affecting the operation due to the inherent delays involved in corruption and expectations. Namely, it takes time for issues or concerns with corruption and expectations to develop. Further, all projects would be routed through the ISF and only vetted contractors would be used to help limit corruption. This was especially important because due to the limited duration of this mission, there would not be time to conduct a drawn out bidding process for projects. Projects would have to be quick and have an immediate impact in order to gain the desired outcome.

After attempting to understand the most basic aspects of the problem and developing guiding principles for the operation, the next major challenge for the Task Force was to analyze the relatively sparse information that previously existed in the area and to develop basic information on the status of critical infrastructure, roads, political, government and tribal institutions, and on basic demographics that would focus reconnaissance operations. This was initially difficult to do as U.S. forces had not been conducting partnered operations in the area at the brigade level or below for several months due to the reduction of forces in Iraq. In addition, the previous units that had operated in the area had long since left Iraq and were not able to be contacted. Nevertheless, one-third AAB and the battalion responsible for the area of operations, 2-7 IN, provided the limited information that was available so that an systems mapping of the problem could be developed (see Fig. 12.3). This process included development of initial target areas of interest (TAI) based on the believed residences of various persons of interest or warranted enemy combatants. These TAIs were then used in developing an execution matrix that synchronized various Task Force assets and enablers in efforts to coordinate and synchronize information collection efforts. There were approximately 212 TAIs, and each TAI of a few square kilometers was further broken down into anywhere from 5 to 30 subordinate areas of interest that would be engaged during patrols (see callout in Fig. 12.3). In addition, these sub-TAIs contained wire diagrams of how various enemy actors were associated or believed to be associated with each other. These TAIs provided an initial systems network depicting what actors were believed to be actively conducting operations against the Government of Iraq south of Baghdad (Fig. 12.4).

Once these steps were taken to understand the problem, the operational concept was finalized. The plan was to move back into the area south of Baghdad where U.S. forces had previously withdrawn and to focus on rebuilding relationships with the Iraqi Security Forces (ISF) and the population so that information could be collected to confirm or deny previously assumed information and atmospherics. In doing so, a number of methods were implemented such as combined medical engagements (CMEs), Humanitarian Assistance (HA) drops, and small reconstruction projects. Despite the many complexities and challenges involved in the situation, the design process showed that if a strong relationship could be built with the Iraqi Brigades and Battalions operating in the area and that the local population perceived U.S. efforts as being philanthropic that both entities would support U.S

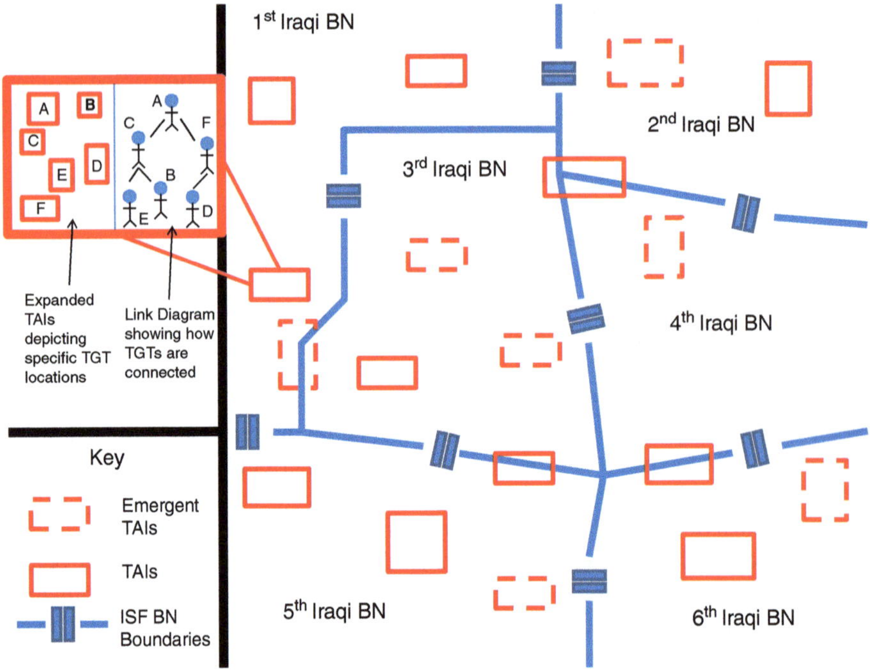

Fig. 12.4 TAI systems mapping

efforts to collect information on governmental allegiances in the area. Consequently, all missions were combined with and approved by Iraqi Security Force commanders so as to avoid undermining their relationship with the community and local leaders and to gain their buy-in.

The HA and CME missions were designed to operate in a "plug-and-play" manner where standing teams and assets could be easily flexed across the more than 5,000 km^2 where the Task Force was operating with minimal notice. For CMEs, a standard package of six vehicles with ten medics, three providers, a pharmacist and eight dedicated interpreters was created. This "CME Platoon" could then move independently between TF 2-14 IN bases and easily be attached and re-attached to different companies so that the medical experts could take part in all steps of the troop leading procedures and develop best practices throughout the operation. It was also critical to the success of these missions to have a balanced mixture of male and female providers as only slightly more women and children came to the CME's than men.

Other enablers such as information gathering teams, social scientists, and tactical PSYOPS detachment teams were integrated into these missions as well in order to collect information and atmospherics, and to disseminate important information operations (IO) messages. Further, by supplying the people with something tangible, it was believed that the Task Force could gain automatic credibility with both the people and the ISF. During the HA and CME missions, conduct of reconnaissance with the ISF prior to execution was critical as this gives the commander on the ground time to integrate all of the assets available into the plan. It was also important to be prepared for contingencies on the day of execution. This is specifically important because the ISF did not always follow our troop leading procedures process and sometimes their manpower was pulled away at the last minute in order to react to the latest threat in the area of operations. By staying flexible while conducting these missions, and always having a good contingency plan, HA and CME missions were believed to be the key to success.

Special emphasis was also placed on building partnerships with the ISF units in the area. One technique that was planned was for TF 2-14 IN to partner down to the lowest units possible. This means that even if a U.S. company (a 130 man organization) was to be partnered with an ISF Brigade (a 2,000 man organization), habitual relationships would need to be established down to where a U.S. platoon (approximately 30 Soldiers) is operating with an Iraqi Battalion (approximately 500 Soldiers), or even better, with an Iraqi company or platoon. The Task Force knew that this would be important not only for establishing best practices and building friendships, but also because it provides better situational awareness for the U.S. force as to what Iraqi forces are actually seeing on the ground. Indeed, a large percentage of the information collected during the mission came from ISF at the battalion level and below and not directly from the local population. In addition, the Task Force planned to use CERP funds, specifically the Commander's Small Scale Project program, to help build a rapport with the ISF as they saw that U.S. Forces in their area were willing to help their people. Small rewards also helped bridge the gap between ISF and USF, as this provided a funding source where Iraqi sources could be paid for actionable information and weapons caches. It was also critical to spend time with ISF partners not just planning and conducting operations, but also relaxing and socializing. Likewise, Key Leader Engagements (KLE) with ISF and Sons of Iraq Leadership was critical to the information sharing within the area. This helped to earn their trust and build a true friendship between Americans and Iraqis quickly so that the mission could be accomplished (Picture 12.2).

The operation was much more successful than anyone expected. The design process was critical in achieving this because it allowed the Battalion to quickly focus in on what aspects of the problem were most important and to develop a quick and simple, but effective plan. Specifically, the operation confirmed and denied various aspects of the TAIs that were initially developed, with the added benefit of providing aid and assistance for thousands of Iraqis and built relationships that will prove important for the strategic partnership between Iraq and the U.S. for many years to come. During the 6 week operation, TF 2-14 IN conducted over 100 combat patrols, in conjunction with its ISF partners killed or captured more than 60 criminals and terrorists, found a dozen caches consisting of various forms of enemy equipment, conducted 20 humanitarian assistance missions and 7 combined medical engagements that positively affected the lives of more than 20,000 Iraqi citizens. The battalion also conducted 8 humanitarian and minor reconstruction projects and produced more than 40 detailed reports of significant information value. The crowning achievement for the entire mission occurred when the Task Force acquired information that led to the identification and detainment of

Picture 12.2 TF 2-14 IN Soldiers conducting CME and HA missions

Latif Adnan Muhammad Hamza Al-Qarghuli, the Al Qaeda leader who ordered the kidnapping, torture and killing of eight soldiers from the Task Force's sister battalion in 2007.

12.5 Heuristics for Solving Complex Problems

Based on lessons learned from these two case studies, the following heuristics are proposed for both private and public sector organizations of all types that are faced with a challenging, complex problem. While all problems are unique, there are certain principles that can be applied to most complex problems to help better understand them and develop effective solutions.

Iterative Design is Key. The design process for solving complex problems is not sequential, but is cyclical. Indeed, once the unity meeting began, the problem continued to be developed as key mavens attended meetings and presented new aspects of the problem to the group. In the case of TF 2-14 IN in the Triangle of Death, the lack of time to conduct detailed planning necessitated that the problem be developed simultaneous to the conduct of operations. This implies that sometimes not only the solution, but the problem itself, cannot be fully understood until execution of an operation or plan. In addition, sometimes the solution or the most important information regarding a problem can come from an unexpected source. In the case of the unity conference, the planning team initially believed that the problem was clearly one of reconciliation. But local Iraqi leaders clarified that the real problem was one of unity, not reconciliation. Coincidentally, constant stake holder input is critical throughout the duration of any project. This is an important point, because often stake holder analysis is considered early in the design process. But, as these case studies show, it is important to revisit stake holder analysis throughout the design process because often times during the beginning of any project, the design team still does not fully understand the problem and therefore, is not asking the right questions. Similarly, constant application of critical thinking where assumptions and even pre-believed facts are questioned in relation to new information is essential to understanding complex problems. Indeed, the challenge here is that complex problems, by nature, can never be fully understood; only better understood. When faced with a complex problem, never assume that your fully understand it.

Think Holistically. Systems Thinking tools are tremendously helpful for understanding complex problems. Both of the case studies provided show how Army planners used various Systems Thinking tools to analyze how various aspects of the problem were inter-related. In the case of the Unity Conference, collecting detailed information on hundreds of key personalities in Eastern Baghdad, analyzing the links between these personalities, and then plotting them using Network Analysis software allowed the team to understand how all key players in the region were inter-related. In the case of TF 2-14 IN's mission in Southern Baghdad, use of a System Dynamics model allowed the team to analyze a multitude of variables that impacted the problem, and more importantly, understand how they were inter-related. As discussed in the Design section of this paper, there are many Systems Thinking tools that a design team can use to help think holistically about the problem. Key to understanding complex problems is choosing the right tool, and going through the process to help the design team think holistically about all aspects of the problem.

Keep an Open Mind. This idea runs parallel to holistic thinking. Maintaining an open mind is critical because it allows the group to constantly assess and re-assess its assumptions throughout the design process. As discussed in the TF 2-14 IN problem, many aspects of the problem emerged as the operation developed. This forced the design team to adjust many of their assumptions during the mission regarding which TAIs were most important. In the case of the Unity problem, keeping an open mind allowed the group to realize that after a month of hard work, a major shift in emphasis and focus was needed. The ability to think holistically and conduct iterative design is dependent on this seemingly simple, but in practice often challenging principle. Avoiding Group Think is critical.

Simplicity. While complex systems diagrams are often useful in efforts to understand complex problems containing thousands of variables, in many cases the solution is simplistic. This is because within any system, often only a few of the variables are responsible for the desired or undesired behavior within the system. For example, there were hundreds of challenges and barriers facing Task Force 2-14 IN in the Triangle of Death, so many that initially some leaders believed that the mission was ill-designed. However, by focusing on the most important aspects of the problem, building relationships with the ISF and with the local population, synergy was created that allowed the Task Force to overcome or bypass the other aspects of the problem. During development of the Unity concept, the number of schisms and dividing factors were so great that focus on them had to be abandoned entirely. Instead the problem was reframed entirely, to focus on how all entities could be brought together rather than focusing on the hundreds of factors that divide them. This may seem simple, but it took the Unity working group almost a month to recognize the value of this important heuristic. Don't assume that the solution has to be complex, just because the problem is.

Listen to your stakeholders. Iraqi mavens provided invaluable input and even provided re-direction to the working group in terms of focusing on Unity in lieu of Reconciliation. Ultimately, the final solution to the Unity group was simply to create a venue where the diverse players could come together and talk about their concerns, fears, challenges, and listen to each other. In this case, the stakeholders were in essence, the solution themselves. In the case of TF 2-14 IN's mission, the entire focus of the operation was placed directly on the stakeholders, the local citizens who lived in the Triangle of Death and the ISF units charged with providing security in the area. Stakeholders often understand the problem better than anyone else, albeit their perspectives on the problem are often narrow in scope. Bringing stakeholders together so that their perspectives can be analyzed and synthesized is often helpful in understanding and solving complex problems.

Focus on "the gap." When presented with a complex problem, it is often difficult to ascertain where to focus. Complex problems contain so many variables and often contain so many feedback loops between variables that they are impossible for the human mind to understand. A good place to begin is with the three questions presented in the beginning of this paper. What is the current state of the environment? What is our desired end state? What are the tensions that are preventing us from transforming the system from its current state to the desired end state? The answer to the third question can be considered "the gap" that must be crossed in order to transform the system from one state to another. It is important to note, however, that this is just a beginning. In the case of the Unity problem over-focusing on the gap initially led the working group down the wrong path, however, without focusing on the tensions in the system the ultimate framework for the problem could not have been developed.

Team Diversity matters. This principle may be the most important of all. Complex problems are so multi-faceted that rarely can one discipline alone provide the insights needed to understand the problem. In the case of the Unity working group, principle players came from the Human Terrain Team, Information Operations, Psychological Operations, the Provincial Reconstruction Team, the Intelligence Section, the Operations Section, and representatives from all maneuver battalions.

In addition, academic and religious experts in the U.S. and Iraq were consulted as were many different Iraqi religious, political, governmental, military, police and tribal leaders. In the case of TF 2-14 IN's mission, key players included operations, intelligence, fire support, logistics, civil military affairs, Air Force tactical air controller teams, aviation teams, attack aviation, human collection teams, human terrain teams, medical experts, explosive ordinance disposal, military working dog teams, military police, contracting officers, local Iraqi leaders from many different organizations, and various Iraqi Security Force leaders. Different actors bring different perspectives to a problem. Synthesis of these perspectives often provides clarity. The challenge is to get diverse actors to work together. But overcoming this challenge is often worth the price. Having mavericks in a group, or simply people who thinking differently, helps to avoid Group Think. It also forces a team to think holistically about various aspects of a problem and to maintain an open mind.

12.6 Conclusion

Not only are complex problems difficult to understand, but approaches towards solving complex problems are still in their academic infancy. More research is needed to better understand the nature of complex problems and to develop practical techniques, not just complex theories, for helping to better understand and address these problems. Indeed, as we look forward into the future the need to confront complex problems head on will prove critical to maintaining our way of life and further advancing the human condition. As researchers continue development of these techniques and theories, practical methods are needed now to address complex problems. The heuristics presented above should be considered by all managers and leaders faced with complex problems, as they are likely to provide some assistance when confronted with many diverse complex problems.

References

Brandes U, Erlebach T (2005) Network analysis: methodological foundations. Springer, New York

Department of the Army (2010) FM 5.0: The operations process. Washington, DC, pp 3-1 thru 3-7

Elder L, Paul R (2006) The Thinker's guide to analytic thinking. Foundation for Critical Thinking, Dillon Beach, pp 3–7, www.criticalthinking.org

Gladwell M (2002) The tipping point: how little things can make a big difference. Back Back Books, New York

Maier MW, Rechtin E (2002) The art of systems architecting, 2nd edn. CRC Press, New York

Parnell GS, Driscoll PJ, Henderson DL (2008) Decision making in systems engineering and management. Wiley, Hoboken

Sterman JD (2000) Business dynamics: systems thinking and modeling for a complex world. Irwin/McGraw Hill, Chicago

Chapter 13
Making Sense of Messy Medical Data Through Constructed Charts of Functional Distances

Lincoln Gray

Abstract This chapter explains a technique called *constructed cartography* that makes maps or diagrams of events that you might not initially consider appropriate for mapping, including various real or imagined manifestations of diseases and our successful and unsuccessful attempts to mitigate those manifestations. The goal is to reveal underlying patterns in messy data. Starting with data in a form where underlying patterns are elusive, a visual summary is constructed. Information is neither created nor destroyed, just transformed into a more useful form. Two fundamental ideas that allow these charts to be constructed – functional distances and multidimensional scaling – are reviewed. This chapter then describes in detail the data that are needed for constructed cartography. A simulation shows that a sample of 35 cases is sufficient to extract 90% of the pattern in a much larger population. Finally, this chapter reviews several diverse instances where the technique has worked well: administrative funding versus successful vaccinations, migrations, metastases, and symptoms of disease. Finally the potential of aligning our attempts to fight diseases as well as our knowledge of how the disease spreads with ways that the disease actually spreads is outlined. Such a test of congruence of efforts can occur at the individual or at population levels. Places where these constructed charts do *not* line up are places for closer inspection, consideration, effort or training.

L. Gray (✉)
Department of Communication Sciences and Disorders,
James Madison University, Harrisonburg, VA, 22807, USA
e-mail: graylc@jmu.edu

S.F. Kovacic and A. Sousa-Poza (eds.), *Managing and Engineering in Complex Situations*, 225
Topics in Safety, Risk, Reliability and Quality 21, DOI 10.1007/978-94-007-5515-4_13,

13.1 Introduction and Overview

13.1.1 Advantages of Constructed Cartography

Multidimensional scaling excels in extracting underlying patterns in variable data. This process is appropriate to reveal what is 'underneath' noisy data – when you struggle to understand the 'bottom line' in a complex set of numbers. In this way it solves 'wicked' problems.

A constructed chart is a result that emerges *from* the data. This process is unbiased in that the data tell us what is important. We do not start by presuming to know the important trends in the data.

"Humans love maps. They love the mapping of any type of information, not strictly geographical. Maps are bursting with information. They transcend the boundaries of language. Maps are good" (Havey et al. 2011). It is often said that a picture is worth a 1,000 words. Information is neither created nor destroyed in the process of making these maps, just transformed from confusing and obfuscated forms into a unified 'vision' of the important trends therein. Once these charts are constructed, interactive visualizations can be created to display the discovered patterns. Several of these interactive visualizations can be seen at http://www.csd.jmu.edu/csdsquared/.

13.1.2 Functional Distances

This chapter includes explanations and examples of the following steps involved in the process of constructed cartography, which include:

1. Measure functional distances between all possible pairs of five or more items for two or more subsets
2. Input these data to multidimensional scaling
3. Statistical evaluation of the result through a neutral model
4. Find best centers for each subset
5. Interpret results

The idea that underlies all of this work is that of *functional* distances. Functional distances are not physical distances. Functional distances are *proximity measures*, similarities between all possible pairs of some set of events. The key is to define carefully the measure of functional distance.

Once functional distances are defined, they are collected into off-diagonal half matrices for input to a statistical technique called multidimensional scaling (MDS).

13.1.2.1 Explanation of Functional Distances Through Metaphors

We use the idea of functional distances frequently in our speech. Suppose we heard that "The transcontinental railroad brought the east and west coasts of America closer together." Of course this is false in the physical definition of distance. Hurricanes might bring the coasts closer together through beach erosion, and global warming will certainly narrow our continent. But the transcontinental railroad merely increased the ease and decreased the cost of coast-to-coast travel. The important idea to be revisited below is that efforts and costs can be important measures of functional distances. It can be instructive when such measures are not congruent with physical distances.

Suppose we heard that "The stock markets in New York and Beijing have never been closer together." This statement happens to be true in the physical definition of distance, because continental drift is decreasing the width of the Pacific Ocean by a few millimeters per year. But most of us would immediately interpret this statement to indicate a functional similarity in the markets – that share prices tend to rise and fall in synchrony at the two widely separated markets. The important idea to be revisited below is that correlations between ratio measures can be used to calculate functional distances.

Finally consider the statement that "Our two senators have never been closer together." This may or may not be true in the physical definition of distances and would likely change every hour of so as the senators moved about their chambers or on junkets throughout the world. Much more relevant is the implied measure of functional similarity – that the votes of the two senators would be similar. This means that if one votes Yeah, the other is unlikely to vote Nay. The exclusive-or (XOR) would commonly be false for the votes of this pair of senators, quantifying their functional similarity. The important idea to be revisited below is that binary data can be used to calculate functional distances, the sum of XORs measuring 'distance'.

13.1.2.2 Calculations of Functional Distance

Creativity is allowed and encouraged in measuring functional distances. As seen in the examples above and below there are many different ways to derive *functional* measures of similarity or difference. These are called proximity measures in the literature of psychophysical scaling. As an example, consider possible differences between bread and bananas. Interval measures occur when differences are meaningful. Such a proximity measure could be calculated as the absolute value of the difference in $/kg (as in Fig. 13.1). Proximities could also be calculated using other functional measures such as calories (similar at ~100/serving for each) or potassium content (different). Binary Euclidian distances (XORs) could be calculated from the

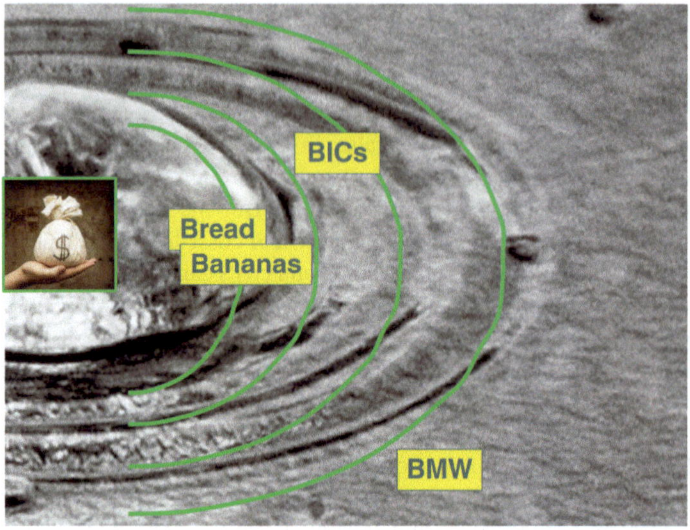

Fig. 13.1 A simple functional model of costs (Background ripples courtesy of Stephen Gill, also seen in Figs. 13.2, 13.3, 13.5, and 13.13)

presence or absence of the two foods in various pantries. Voluntary participants could simply be asked the extent to which they perceived these two foods to be similar or different – a directly measured perceptual proximity. Other indirect perceptual proximities can be derived from confusions or reaction times.

13.1.3 Six Examples of Meaningful Functional Distances

13.1.3.1 Expenses

Constructed charts are easily made from financial data. Money is a ratio scale (having a real zero). An advantage is that these data are often already in a machine-readable form, ready for input to MDS. It is possible to imagine a landscape of our expenses, with costs expanding as 'shock waves' from some starting place into the constructed landscape. In such a model, costs accrue as the shock wave expands. Items that are close together in the chart have similar costs. Imagine bread and bananas as adjacent points versus BICs (disposable pens) and BMWs (expensive autos) as widely separated points. Figure 13.1 shows this model. This idea of expanding shock waves is useful in the other examples.

13.1.3.2 Various Forms of Effort

Funds expended are but one form of effort. Time spent is another form of effort. Among medical endeavors to be discussed later, risks, side-effects, and morbidity

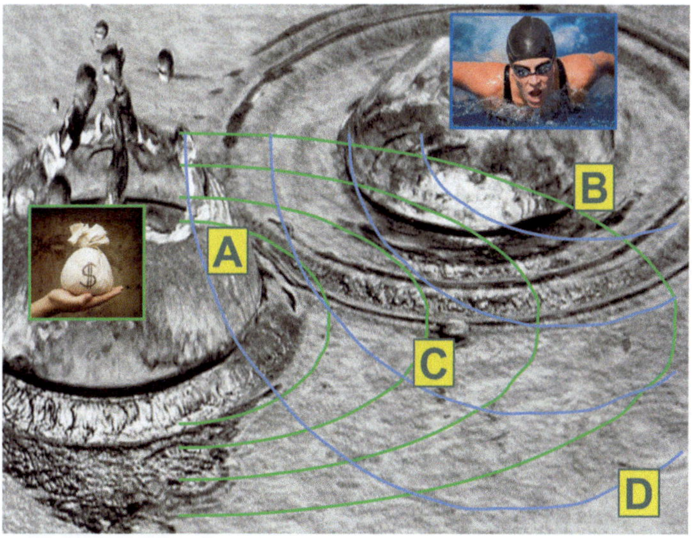

Fig. 13.2 A functional model of two forms of effort

could be construed as independent efforts toward (or away from) a cure. Figure 13.2 shows a metaphor of discrepant patterns in two different forms of effort. The 'overall landscape' of the goals (shown as lettered squares) is the same for both forms of effort, but there are different 'centers of effort'. Expenses and effort (depicted by the swimmer) accrue in proportion to distance from each separate center. Thus, not much money but great effort is spent on item A; lots of money but not much effort on item B, and much money and effort on item D. The two forms of effort are congruent for items C and D but not A and B.

Note that the expanding 'shock waves' modeling the two forms of effort (shown in Fig. 13.2 in green and blue) are of different shapes. This idea will be discussed below as *weighted scaling*.

13.1.3.3 Success

Various forms of measured successes (or failures) can be used to calculate functional distances. A recently published paper on patterns of successful vaccinations in developing countries (Gray et al. 2010) will be reviewed later. Student achievements can be mapped (Gray 2001).

Figure 13.3 shows two overlaid constructed charts where effort in terms of expenses leads to success. The 'shock waves' from both sources of functional distances (measures of effort and of success) emerge from the same position and spread into the same functional topography of items (lettered squares) in the same pattern. This is the desired situation. The congruence of the two maps is quantified through a cannonical correlation, the multivariate extension of the Pearson product-moment correlation (r).

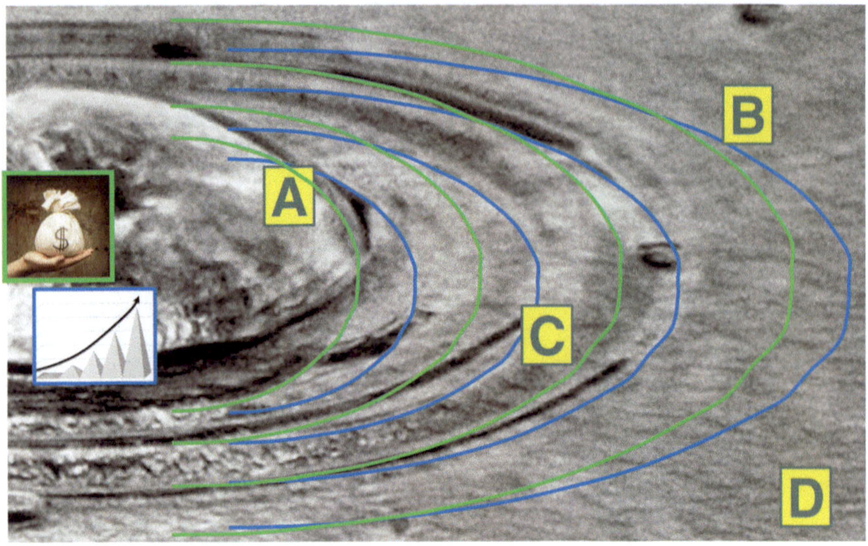

Fig. 13.3 A functional model where success is proportional to effort

Alas, the amount of effort put into an endeavor does not always equate to success. Constructed cartography can help solve this 'wicked' problem of aligning effort with achievement. Constructed charts can be an unbiased assessment of administrative efforts. Discrepancies indicate an impediment – where efforts did not have their intended purpose. These would be places for increased scrutiny for local factors that impede the intended progress.

13.1.3.4 Metastases

Metastases are an interesting example of medical symptoms that illustrate the use of constructed cartography. First consider cancers that spread to the next nearest area. Basal cell carcinomas are a common cancer of sun-exposed pale skin. Although the disease is termed malignant, it rarely kills because these cancers simply spread to the next nearest area. Surgical cure involves excising ever widening margins until the shaved tissue no longer contains cancerous cells. The disease rarely recurs. If a chart of basal cell carcinomas were constructed, it would resemble local anatomy as we see it.

Melanomas, in contrast, are much more deadly skin cancers because they have a tendency to disperse widely. We can, however, construct an imaginary anatomy such that the disease does spread to the next nearest area. Thus, we have taken a disease that spreads in complex patterns in real anatomical dimensions and mathematically transformed it into a disease that spreads in simple patterns to the next nearest area in a landscape constructed by the computer. That constructed anatomy will look strange to us. It is how the disease 'views' our body, not how we view our

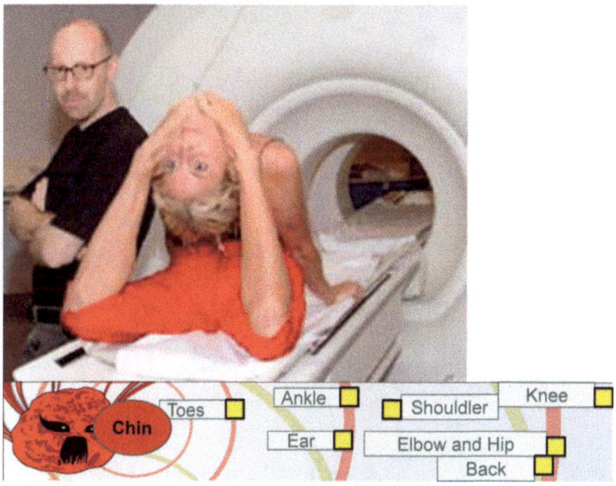

Fig. 13.4 Physically contorted anatomy (*top left*, reprinted with permission from Professor Richard Wiseman, Psychology Department, University of Hertfordshire) and functionally contorted anatomy – a constructed chart of metastases (*bottom*, drawn by Josh See, Creative Services, JMU)

body For diseases that do not spread to the next nearest area, the constructed map will not resemble a standard anatomical map. This is just another statement that the disease does not spread to the next nearest area.

In real anatomy the ankle bone is connected to the foot bone, etc. As a fanciful example, suppose there was a chin-bone cancer that tended to spread to the toe bones. We might imagine some constructed, 'contortionist' anatomy such that the chin and toes were close together, mathematically depicting the probability that these two physically separated areas commonly share disease. Figure 13.4 shows an image of such contorted anatomy. At the bottom of Fig. 13.4 is a constructed chart that might arise from the MDS analysis of such a cancer. At the left edge of the 'map' on the bottom of the figure, we see a 'starting place', a malignant cell 'viewing its options' for possible new locations to reside. Shock waves emerge from that starting point into a functional landscape where distances are proportional to the probability of sharing disease. Cancer from the chin soon spreads to the toes, then to the ear and ankle, then shoulder, then elbow and hip. The knee is the farthest away from the foot, indicating an unlikely pattern of spread in that 'direction'. This is a very strange anatomy, but such a map would explain the behavior of this imaginary disease.

13.1.3.5 Signs and Symptoms of Disease

Metastases are but an example of medical symptoms. Constructed cartography works equally well with symptoms that might not be so easily located. Examples might be the all-too-common medical problems of our society: obesity, high blood

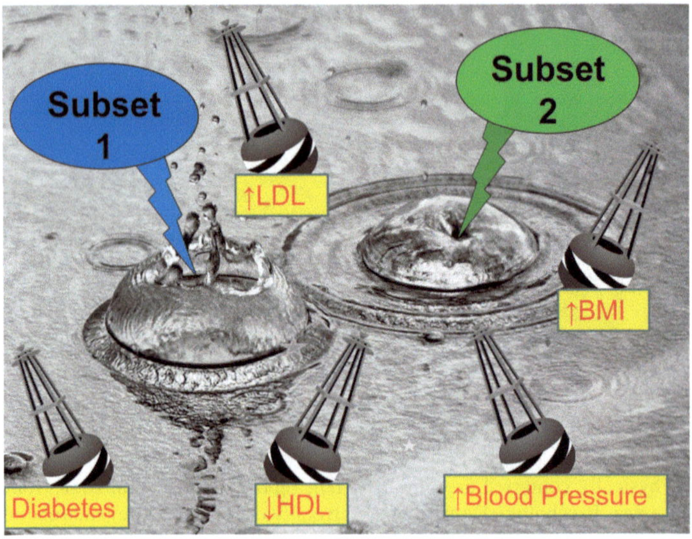

Fig. 13.5 A metaphor of charted symptoms (Buoys by Josh See)

pressure, high cholesterol, diabetes. A useful analogy is that of buoys floating upon a fluid surface. The start of a disease process is modeled as a projectile hitting the surface and thus starting an expanding ripple. A symptom appears (reaches clinical threshold) when the ripple rocks the buoy. Actually the process is reversed: multidimensional scaling finds a topography of symptoms such that such a model adequately represents the raw data. Such a model (or result from the process of constructed cartography) is seen in Fig. 13.5. As discussed below under *weighted scaling* there are two subsets of patients, perhaps different ages or genders that have different patterns of spread.

13.1.3.6 Perceived Risks

MDS was originally developed to map perceptions. Participants could be asked to produce proximity measures by moving a slider to indicate perceived similarity as all possible pairs of a set of items are displayed. A computer program would ask "How similar or different are these items?" (If readers are interested, the links at the end of this chapter offer opportunities to participate in such research). Figure 13.6 shows how such data might be collected. At the top is what the computer might display for one pair of risks (smoking and obesity). Other risks, including alcohol, auto accidents, drugs outdoor adventures, and trans fats, would appear in different pairs. All possible pairs of these items, in this example 21 pairs of 7 items, would appear randomized in sequence and position. The participant would move a slider to indicate their perceived proximity. For weighted scaling there would need to be two of more subsets of participants

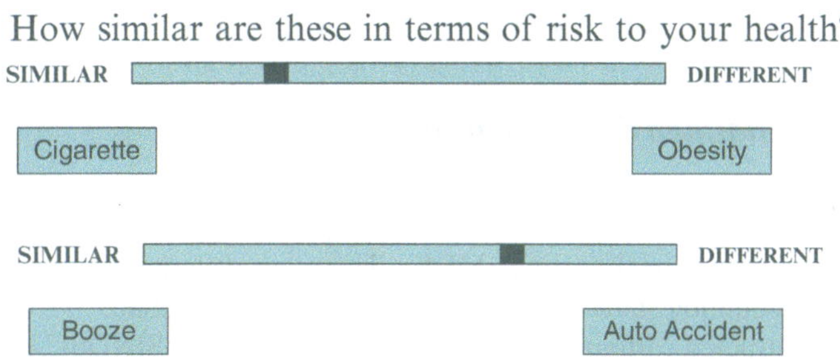

Fig. 13.6 Tests of risk perceptions

Fig. 13.7 A chart of expanding knowledge (By JMU Creative Services)

such as teens and adults, expecting that teens would perceive themselves to be without much risk.

There are several reasons to make perceptual maps of health-care issues (Gray et al. 2002; Herrin and Gray 1997). There is a view that we should all take more responsibility for our health. Rather than expecting physicians to just make us better, a controversial approach might be for health care to be collaboration between professionals and the owner of the body. It would then be desirable for us all to know our own perceptions of various risks and benefits of all that is under our control (eating no salt oatmeal, regular exercise, eschew pastries, etc.) It is likely that perceived risks drive behaviors. An interesting perceptual study was conducted on the perceptions of smokers versus non-smokers on the risk of oral cancers (Herrin and Gray 1997). There was no difference between the perceptions of the two groups, quite different than the reality of medical facts. Interestingly, when the participants were asked to imagine that their smoking status had changed – that smokers had quit and non-smokers started smoking – the imagined perceptions were then very different. This suggests that our imagined perceptions of what would happen if we did something differently probably drives our behaviors more than our perceptions of current risks.

Risk perceptions are closely related to knowledge maps. We could imagine maps of our knowledge about various diseases. Figure 13.7 is a simplistic image of how knowledge accrues over a lifetime. More relevant, but conceptually similar would be maps of expanding knowledge about the current wicked problems in our society: HIV disease, metabolic syndrome, vascular disease.

13.2 The Process of Constructed Cartography

13.2.1 Multidimensional Scaling

The underlying 'engine' of constructed cartography process is the statistical technique of multidimensional scaling (MDS). There are different versions of MDS found in most common statistical packages (SPSS, Systat, Matlab, …). Particularly useful for constructed cartography is an older version, called SINDSCAL for symmetric individual differences scaling (Pruzansky 1983). The general idea of MDS is first reviewed below, then a major subset of MDS algorithms called weighted, individual-differences-, or three-way-scaling is described.

13.2.1.1 General MDS

MDS is easier to describe in the reverse. Suppose we had a map of a set of items, perhaps five large cities in the 'corners' of America as seen in Fig. 13.8 (note that physical distances are, perhaps unfortunately, used in this initial illustration of a process that will elsewhere use functional, not physical distances). Given such a map it would be easy to construct a matrix of distances between all possible pairs of cities. Such a matrix is frequently found at the bottom of travel maps. Only ten distances are necessary, even though a 5 × 5 matrix would have 25 cells. We assume that the distance from each city to itself is zero, so the diagonal elements are not needed. We further assume that the distance from city A to B is the same as the distance from city B to A. Thus the matrix is symmetrical. All that is needed is the off-diagonal half-matrix of distances. For a map of N points the off-diagonal half matrix would contain what mathematicians write as ${}_{n}C_{2}$, the number of possible pairs of N items. This equals N*(N – 1)/2 (Table 13.1).

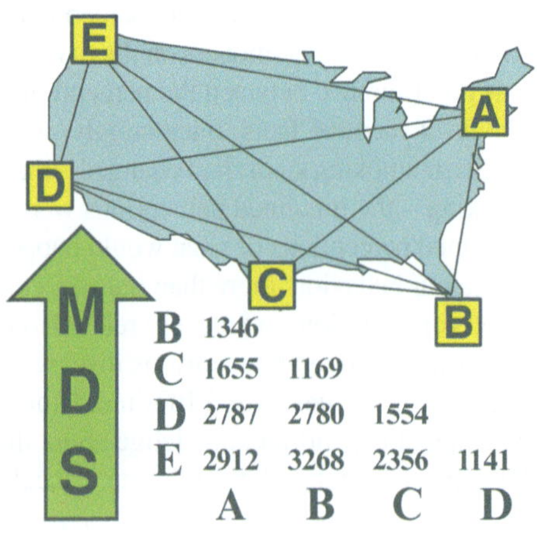

Fig. 13.8 MDS of intercity distances

Table 13.1 ${}_{n}C_{2}$ for various n (number of possible pairs for different numbers of items)

Items	5	7	10	20	50
Pairs	10	21	45	190	1,225

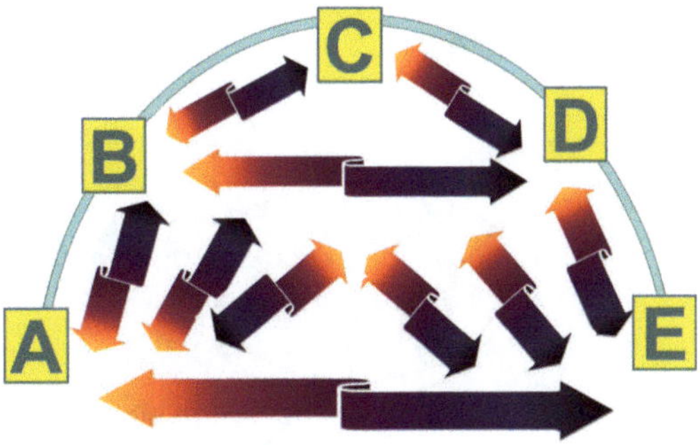

Fig. 13.9 A constructed chart 'squeezed' from functional distances

MDS takes the off-diagonal matrix and from it constructs a map in some stated number of dimensions such that distances in the map are maximally correlated with the input data (Kruskal and Wish 1978; Schiffman et al. 1981). If the data were those from the five cities in America, MDS would produce a plot of these points roughly as we see in common wall maps. MDS also returns a goodness-of-fit, an indication of how well the result fits the input data. If we were to request a two-dimensional map of the five cities, then the fit would be good but not perfect, because America is three-dimensional.

Another way to think of the process of MDS is shown in Fig. 13.9. Ten different proximity measures are shown as partially folded ribbons (or they could be springs). These ribbons represent all possible comparisons between five items shown as the squares labeled A–E. The length of the ribbon represents the magnitude of the functional distance that was measured: longer ribbons for large distances. Think of these ribbons as labeled on each end to represent the two items that were compared. There would be four ends labeled A that would have B–E on the other end; four ends labeled E that would have A–D on the other end, etc. Now suppose that all of the ends labeled A were tied together, all the ends labeled B tied together, etc. What MDS does is find positions of these five knots (A–E) such that the ribbons are perturbed (shortened or lengthened) as little as possible.

From this analogy it can be seen that decreasing the dimensions, such as forcing the ribbons in Fig. 13.9 into 1-dimension would decrease the goodness-of-fit, that is increase the perturbations or 'stress' in the lengths of the ribbons necessary to draw them into a narrow tube – the analogy of MDS solution on a line.

13.2.1.2 Weighted Multidimensional Scaling

A powerful version of MDS includes individual weights (Pruzansky 1983; Schiffman et al. 1981). An explanatory metaphor is seen in Fig. 13.10. Suppose that functional distances between cities were measured not in miles but in days of travel time in

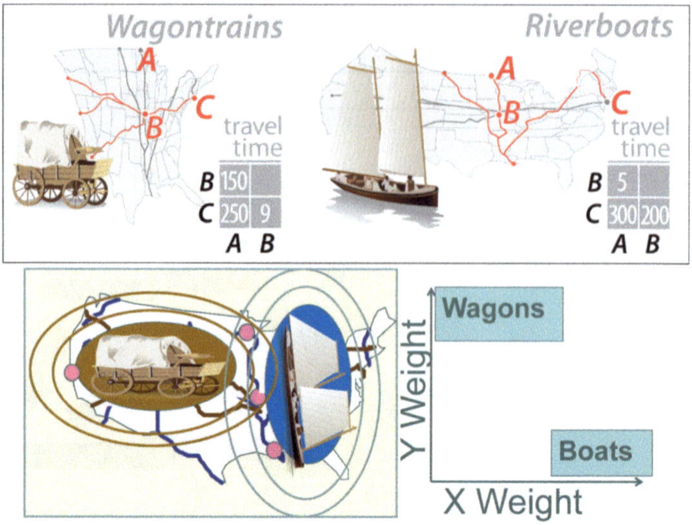

Fig. 13.10 A metaphor of weighted functional distances (By JMU Creative Services)

eighteenth-century America. Suppose further that travel was by either wagon-trains or river-boats, and that rivers flowed north/south while wagon trails went east/west. Minneapolis would be 'close' to St. Louis by boat but 'very distant' by wagon. Similarly St. Louis could be reached from Washington, DC by wagon on the Lewis and Clark Trail but not by boat. Two off-diagonal half matrices of these functional distances measured from two sources (boats and wagons) could be input to weighted MDS. Three results are obtained: the best fitting map and a goodness of fit (just as in regular MDS), but now a set of individual weights are also calculated. It is assumed that the overall map is the same for all sources but that that map can be stretched to best fit the data from each source. We can imagine that north/south distances would seem shorter to a boat captain than to a trail-boss and vice versa. These individual weights can be depicted in three different but mathematically equivalent ways. These three depictions are seen in the three panels of Fig. 13.10. The constructed chart can be stretched differently for the two sources. This is seen in the top panel with two different versions of America: a horizontally stretched version for boats on the right and a vertically stretched version for wagons on the left. This relationship between the boats and wagons can equivalently be depicted as ellipses with different axes and orientations expanding into a single map. This is seen in the bottom left panel with dashed, horizontal ellipses for the wagons and solid, vertical ellipses for the boats. Finally the traditional presentation of individual weights is shown in the bottom right. This is a plot of the weights that are output by SINDSCAL or other weighted-scaling algorithm. Often it is useful to plot an angle for each subset; this would be the arctangent of the Y-weight/X-Weight. The norm of the weight vectors, $(\text{XWeight}^2+\text{YWeight}^2)^{-2}$, indicates how well the data from each source (or subset) fits the overall map. Note the orthogonal orientations of the maps and the expanding ellipses by considering the view of a boat captain: the horizontal extent of America would seem insurmountable, the X-Weight is very

Table 13.2 Organization of raw numerical data for constructed cartography

	Items to be mapped (costs, counts, times, …)				
Meaningful subsets	1	2	…	5	…
a	#	#	#	#	#
b	#	#	#	#	#
c	#	#	#	#	#
…	#	#	#	#	#

high, and boats would spread easily in the vertical dimension. The 'view' of a trail-boss would be exactly opposite. This idea of stretched functional distances for different types of effort or for different subsets of data is seen in various examples in Figs. 13.2, 13.3, 13.4, 13.5, and 13.11, and for different forms of real medical data in Figs. 13.13, 13.16 and 13.17.

13.2.2 The Input Data Needed for MDS

As indicated above, many different forms of data can be collected and then transformed in various ways to form the proximity data necessary for multidimensional scaling. The minimum data for a two dimensional constructed chart are measures in about five columns for three or more subsets, though more data are often needed for simple tables of ratio-level data (option 1, below).

13.2.2.1 Ratio-Level Data

Table 13.2 depicts appropriate interval or ratio scale raw data for constructed cartography. The items to be mapped could be many of the examples reviewed above: costs, numbers of people with some symptoms, numbers of persons effectively immunized, travel times, etc. For weighted scaling there must be two or more meaningful subsets. These subsets could be different genders, data from different countries, or people with two different varieties of the disease (or travelers on wagon trains or river boats in the fanciful illustration). Proximity measures can be calculated as the absolute value of the difference between all possible pairs of numbers in each row.

13.2.2.2 Binary Data

Binary data are 1s and 0s, indicating the presence or absence of some set of traits. The rows are individual cases (such as patients) in two or more subsets. Any symbols (such as +, –, and ?; or 1, 0, and blank) can be used to indicate if the trait is present, not present or unknown. When assembling such data, other columns could be added for any data that might indicate different subsets. It is often advantageous to collect interval data, even if the analysis will eventually be done on binary

Table 13.3 Organization of raw binary data for constructed cartography

Primary		Items to be mapped				
Subset	Case	1	2	…	5	…
a	1	+	−		−	
	2	+	+	−	−	
	…					
	N_a					
b	1	+			+	
	…					
	N_b	−	−	+		

data. In medical data there are typically measures that could then be transformed to binary results based on established norms. For example, rather than a simple binary entry for obesity, weight and height could be recorded and then transformed to a binary entry if the BMI > 24.9. The binary Euclidean distance is a good measure of proximity from such binary data. This is the square root of the sum of the *exclusive ors* between two columns of 1s and 0s. The exclusive or (XOR) is true if both items are different: XOR (1,1) = 0, XOR (0,0) = 0, XOR(1,0) = 1, and XOR(0,1) = 1. Thus, the XOR is a measure of functional distance. Pairs of items with many true XORs are different. For each subset, the binary Euclidian distance is calculated for all possible pairs of items (Table 13.3).

Missing data are typically treated as 0s. Constructed cartography, however, provides a powerful method to analyze patterns in missing data. Charts can be constructed from XORs calculated in two different ways: first where 1 = present and 0 = absent or missing, and second where 1 = measured (either present or absent) and 0 = not measured. If the two charts are congruent (significant canonical correlation) then the human efforts to collect the data are appropriate for the pattern of spreading disease. If the constructed charts are different, then human behaviors are not aligned with the disease's behaviors.

13.2.2.3 Directly Measured Proximities

A third option is directly measured proximities. In an example shown in Table 13.4, there are two subsets and five items to be mapped. The proximities (Δ) are organized into off-diagonal half matrices ready for input to MDS. There are ten pairs of five items: ten entries for subset a (Δ_{rca}) and ten for subset b (Δ_{rcb}). In this notation the first subscript is the row item, the 2nd subscript the column item, and the 3rd subscript is the subset.

A useful but indirect method to collect matrices of proximities is to measure reaction times as the items are changed from one to another. This was used in a study mapping the perceptions of phonemes in children at risk of reading disability (Breier et al. 2004). A computer repetitively presented one sound (/Ga/, /Ga/, /Ga/ …) and then changed the token maintaining the ongoing rhythm (/Ga/, /Ga/, /Ka/, /Ka/ …).

Table 13.4 Organization of proximity data for input to MDS

Δ_{21a}			
Δ_{31a}	Δ_{32a}		
Δ_{41a}	Δ_{42a}	Δ_{43a}	
Δ_{51a}	Δ_{52a}	Δ_{53a}	Δ_{54a}
Δ_{21b}			
Δ_{31b}	Δ_{32b}		
Δ_{41b}	Δ_{42b}	Δ_{43b}	
Δ_{51b}	Δ_{52b}	Δ_{53b}	Δ_{54b}

Table 13.5 A 5×5 latin square

a	b	c	d	e
1	2	3	4	5
2	3	5	1	4
3	5	4	2	1
4	1	2	5	3
5	4	1	3	2

Participants were instructed to press a button as soon as they heard a change. Odd numbers of items are needed in order to present all possible transitions without unnecessary repetitions. For example, five items (a–e) could be presented in the order a-b-c-d-e-a-c-e-b-d-a to present all ten possible transitions (${}_5C_2 = 10$). Similarly seven items (a–g) could be presented in the order a-b-c-d-e-f-g-a-c-e-g-b-d-f-a-d-g-c-f-b-e-a to present all 21 possible transitions (${}_7C_2 = 21$). Latin squares provide an unbiased method for balancing the orders of such a presentation of transitions. A Latin square has one and only one item in each row and column. Table 13.5 shows one of the 161,280 possible 5-by-5 Latin squares. To design an experiment, participants would be assigned to a condition specified by one of the rows, assigning the five different signals (1–5) to the ordered element (a–e) in the sequence of ten described above.

13.2.3 How Much Data Are Enough?

Power-analysis/sample-size techniques have not yet been developed for MDS. Thus, multiple simulations were run, randomly selecting subsets of different sizes from a larger study (Gray et al. 1993) and comparing the MDS solution from the smaller subsets with the result from the larger number of cases. Ten different samples of 10, 20, 30 and 40 cases each were randomly selected from a set of 1,400 cases and subjected to MDS. The correlation of the resulting map with the map from all the cases was calculated using Wilk's lambda from the canonical correlation of the two, two-dimensional maps, and the *worst* result in ten random samples was used to

Table 13.6 A power analysis for constructed charts: the goodness-of-fit expected from various sample sizes

Sample size	VAF (1–λ) (%)
10	73
20	82
30	87
40	92

indicate the power of the analysis. Wilk's lambda (λ) shows how much variance is *not* explained by the multivariate correlation between the two maps. Thus, $1 - \lambda$, in Table 13.6 show the variance-accounted-for (VAF) by each small sample. There is better than a 90% chance of achieving 90% of the desired result from as few as about 35 cases.

As a rough guide to how much data is desirable, statistically significant results have been obtained from 9 subsets and 18 items in as few as 32 (above) or as many as 1,400 cases (Gray et al. 1993), as have 3 subsets and 8 items in 290 patients (esophageal cancers, unpublished), as have 8 subsets and 8 items (cranial nerves) in 95 patients (Gray 1997), as have 5 subsets and 3 items (breast cancers) in 150 cases (Gray et al. 2004), as have 8 students judging each others' performance (Gray 2001).

13.2.4 How Many Dimensions?

Multidimensional solutions (the functional maps) can be constructed in any number of dimensions. Goodness of fit always increases as the number of dimensions increase. Guidelines for the appropriate number of dimensions are to find the number of dimensions just after the last big jump in goodness of fit, to find an interpretable result, and to find a result that is statistically significant.

13.2.5 Neutral Model to Establish Statistical Significance

There are no standard tests for the statistical significance of an MDS result. But a fundamental question for any result is 'what is the probability of getting a result this good or better purely by chance?' A neutral model (also known as a Monte-Carlo or jackknife procedure) can be used to estimate this probability. This brute-force method is always appropriate because it makes no assumption about the underlying distribution of results. The goodness-of-fit of the MDS result from the real data is taken as the important result for the purposes of this statistical evaluation. Sets of data, similar to the real (biological) data, only randomly generated or shuffled, are repeatedly subjected to the same analysis as done for to the real data. For each iteration of the random process the resulting goodness-of-fit is retained. After many

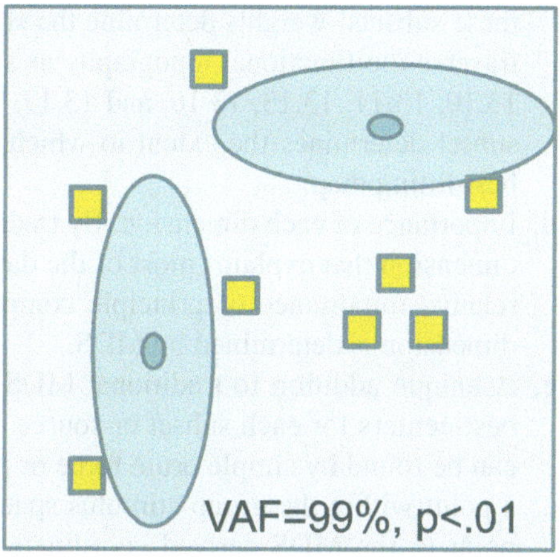

Fig. 13.11 Depiction of MDS results in constructed charts

iterations of the random simulations, a histogram of these random goodness-of-fits is accumulated, and it is easy to determine if the goodness-of-fit from the real analysis is above the level achieved in the 95th best fit in 100 random simulations. If so, then $p<.05$ can be reported.

Careful matching of the distribution of the raw data seems not to be very important. Several simulations were run matching the distribution of interval data to a beta distribution (estimating two parameters from the input data) and then randomly drawing numbers from such a distribution. The conclusion of whether the result was statistically significant at the .05 level was not different when the entries for the neutral model were simply taken from a uniform distribution (0–1). What is important is to have the same number of items and subsets as in the analysis of the real data. When modeling binary data, use a percentage of random 1s and 0s matching that in the real data, and then calculate binary Euclidean distances (sum XORs) for repeated simulations.

13.2.6 Interpretation of a Constructed Chart

The output of MDS and the formation of a constructed chart occur in several forms:

a. Most important is the overall best-fitting map for all of the data. This is called the group-stimulus space in the literature of MDS. The squares in Figs. 13.1, 13.2, 13.3, 13.4, 13.5, 13.8, 13.9, 13.11, 13.13, 13.15, and 13.18 are examples of group-stimulus spaces.
b. The fit of that best-fitting map to the input data. This is a correlation coefficient in SINDSCAL (Pruzansky 1983). R squared is the variance-accounted-for (VAF), and can be used as an effect size. This VAF can be compared to results from the neutral model to derive a statistical significance for the results.
c. The subsets' weights. In the literature of MDS these are called individual weights, because in psychology the separate matrices of functional distances come from different individuals. A contribution of the present chapter is to suggest that many other different subsets can produce instructive results. In the present analysis

these subsets' weights determine the shape of the ellipses (or shock waves) that traverse the functional topography as seen in Figs. 13.1, 13.2, 13.3, 13.4, 13.5, 13.10, 13.11, 13.13, 13.16, and 13.17. The norm of the weight vectors for each subset determines the extent to which the data from that subset fit the overall best-fitting map.

d. Importance of each dimension. By tradition the X (horizontal) axis is always the dimension that explains most of the data. Like the eigenvalues that describe the relative importance of principle components, the relative importance of each dimension is determined by MDS.

e. A unique addition to traditional MDS described in this chapter is finding the best-centers for each subset or source of the functional distances. These points can be found by simple brute force or by elegant optimization. For each subset, a point within the group-stimulus space is found such that distances from that point to the MDS-derived coordinates is minimally correlated with the total amounts for each item for that subset. We seek large negative correlations because close distances mean more of whatever item is measured. For the ratio-level data described in B1 above, those totals would be the numbers in the table. For the binary data described in B2, those totals would be the sums of 1s in each column, added separately for each subset. Best centers cannot be found from the direct measures of proximity, described in B3 above. A separate query of the participants is necessary to derive measures needed to find best centers. This could be a slider for each item under some prompt that specifies a uni-dimensional trait such as "How important is this item?" Another very powerful technique is to use radio buttons under a repetition of al possible pairs asking "which of these items has more of some trait?" (Gray 2001; Gray et al. 2002; Herrin and Gray 1997).

13.3 Successes in Constructed Cartography

Several successful implementations of constructed cartography are now described becoming progressively more abstract in the measures of functional distance.

13.3.1 Migrations

A map of various countries was made, considering the probability of immigration as the important measure of functional distance (Noftsinger et al. 2008). Figure 13.12 shows a cartoon of this process. If there was much immigration between New York and Sydney, then these two geographically distant cities would be considered functionally adjacent.

Fig. 13.12 Functional patterns of migration (By JMU Creative Services)

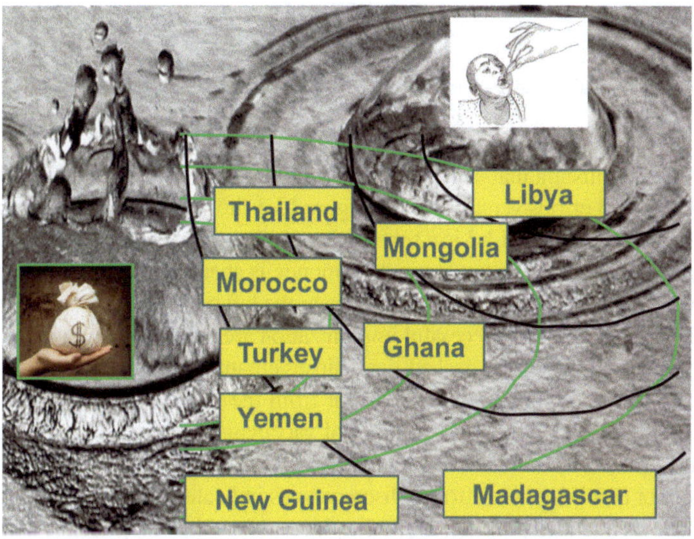

Fig. 13.13 A functional map of vaccinations, where effort does not equal success (Picture of polio vaccination reprinted with permission from Rod Shaw, WEDC, Loughborough University, originally produced for IFRC, Geneva)

13.3.2 Vaccinations

A recent publication charted successful immunizations throughout the developing world (Gray et al. 2010). Different countries are points in the map; starting places and shock waves are properties of vaccines. As in the example above, shock waves expand across a 'foreign landscape'. Like concentric bulls-eyes in a dart game, the closer a country is to the starting-place of a vaccine, the more likely are effective immunizations in that country. Different immunizations (measles, TB, ...) and different ages (children and adults) have different starting places, meaning that the patterns of success are different. Figure 13.13 is a subset of the data in the published analysis. The only immunization shown is that for polio, although there were seven vaccinations in the original analysis and only 9 of the 47 countries are plotted.

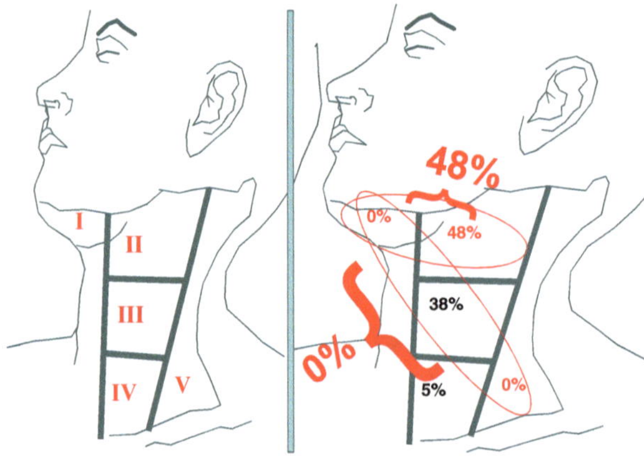

Fig. 13.14 Functional distance is not physical distance in metastases to the neck

An important insight that can arise from a comparison of multiple sources of data is the comparison of efforts versus success. Figure 13.3 shows an example of a chart where success equals effort, Fig. 13.2 is a schematic of a discrepancy, and Fig. 13.13 shows a result from actual data. The pattern in how money is spent by different governments is shown by the solid-line ellipses emanating from the left edge of the constructed chart in Fig. 13.13. The patterns of successful immunization are *not* congruent with the patterns of money spent, shown by the dashed-line ellipses emanating from the top. For example, compare Libya where there is little cost (far from the 'bank' in the figure) and much success (close to the 'bull's eye' or 'center of vaccination'), versus Yemen where there is more cost (closer to the bank) and less success. Constructed charts can be an unbiased assessment of administrative efforts. Discrepancies indicate some impediment – money did not have its intended purpose. Thus, the places where the mapped shock waves from different sources of functional distances do *not* match are places for increased scrutiny for local factors that impede the intended progress.

13.3.3 Metastases

Several successful charts of metastases have been constructed and described (Cousan 2004; Gray 1995,1996; Gray et al. 1992, 1993, 2000, 2004). The 'interesting' charts are of diseases that do not spread to the next nearest area. Most of this work has been done with oral (head and neck) cancers. Head and neck cancers have an unfortunate tendency to metastasize to lymph nodes throughout the neck. Figure 13.14 shows the physical locations to which they spread (neck levels I–V) as well as an indication of how functional distances can be different than physical distances. In the right panel we see some data from oral cancers. Although Levels I and II are physically adjacent,

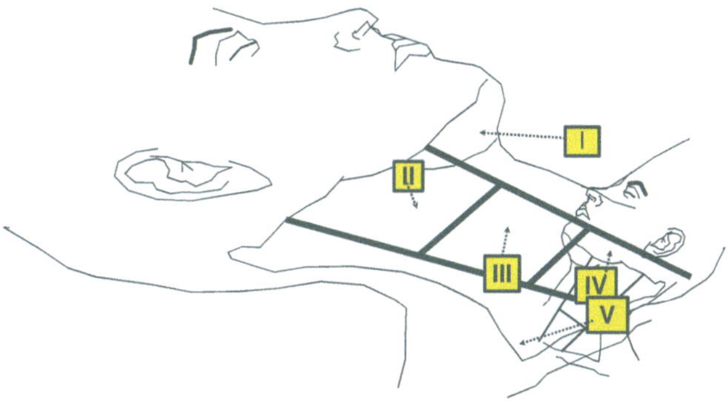

Fig. 13.15 Functional and physical maps of the neck

they have very different probabilities (48% difference) of attracting metastases, and so their functional distance is large. Levels 1 and V, in contrast, are far apart in the neck, yet they have identical properties in terms of metastases from these tumors and so they would have low (zero) functional distance.

Figure 13.15 shows the MDS result – the constructed chart of the neck levels superimposed on rotated and stretched physical sketches of the neck. The nodes on the opposite side of the neck are clustered in a shrunk and twisted mirror image in one corner of the constructed chart.

A fascinating result appears in the analysis from weighted scaling. Here the different subsets were different places where these tumors could start: floor or mouth, tongue, ... larynx. Figure 13.16 shows the plot of these weights in the left panel. This is in the same form as the plot of the weights in the bottom right panel of Fig. 13.10. The right panel of Fig. 13.16 shows the physical locations of these starting places in adults, the subsets that formed the weighted scaling analysis. All of the data came from adults, since oral cancer is a disease that arises after many years of smoking and drinking. It is interesting to speculate that something in embryology might influence future cancers' behaviors, as the configuration of the points in the early embryo resembles the left panel, and this configuration matures to the configuration seen on the right.

As elaborated above, these weights can be taken as shapes of elliptical waves that expand into the constructed chart. This sets up an opportunity to make interactive visualizations of the predicted processes. As the waves expand they encounter more and more of the places that might harbor metastases. This probability of disease can be visualized as changing colors on a realistic image of physical dimensions, seen in the small inset of Fig. 13.17. This interactive, three-dimensional depiction ran on a Cray supercomputer (Gray et al. 1993). Users could 'spread' the cancer and observe predicted metastases on a realistic transparent head that could be rotated to view the 'hot spots'. A two-dimensional visualization can be seen at http://www.csd.jmu.edu/csdsquared/apps/rinc1/rincApplet.htm.

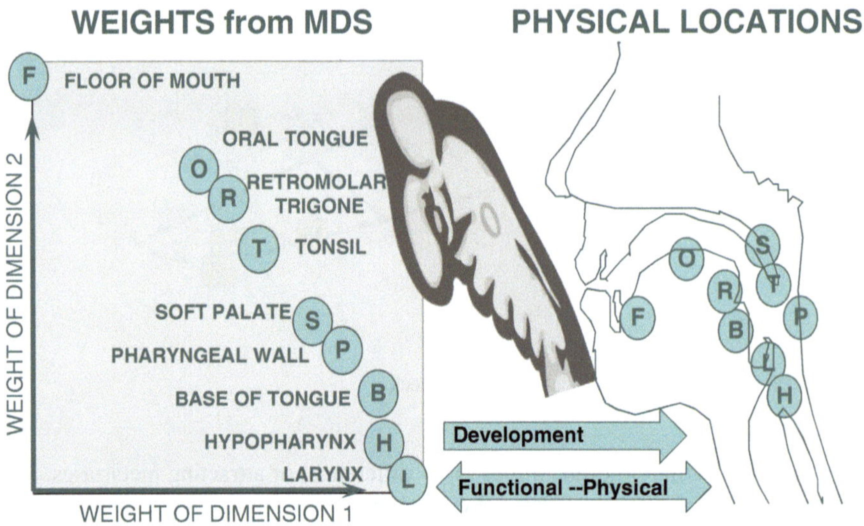

Fig. 13.16 Embryology in a constructed chart of metastases. The primary weights shown in the *left panel* fit the positions of these structures in early development (Adapted from Gray et al. 1993, reprinted with permission from Sage; sketch of early human embryo by Josh See)

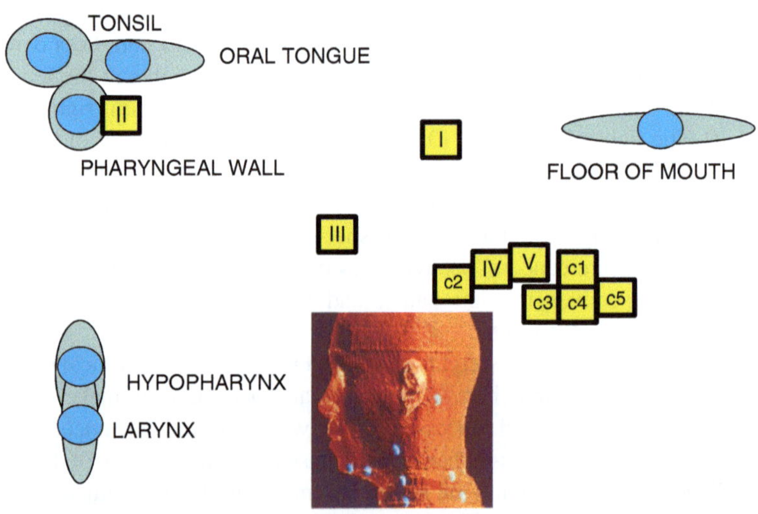

Fig. 13.17 Visualizing predicted metastases. A simplified version of the functional map from Gray et al. 1993 is depicted with an *inset* that was used to interactively display the model in physical coordinates. Contralateral nodes are in smaller squares labeled cN, where N is the Arabic number of the neck levels I–V

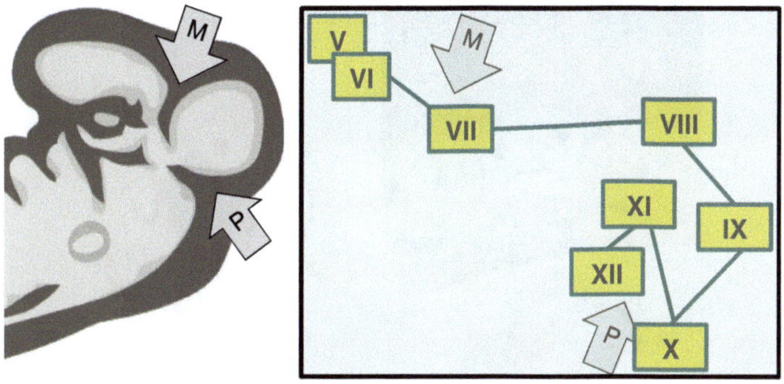

Fig. 13.18 A pattern from embryology in a constructed chart of clinical symptoms. A sketch of the early human embryo is seen in the *left panel*, and the MDS result from a mapping of neural tumors in adult patients is seen on the *right*. Note the similar depictions of the mesencephalic and pontine flexures, pointed to by the *arrows* labeled M and P respectively in both panels (Sketch of early human embryo by Josh See)

13.3.4 Symptoms

Metastases are but one example of a medical symptom. Metastases have an actual physical location, but a chart could be constructed of symptoms that do not have a location. This was done with symptoms from some tumors of our cranial nerves (Gray 1997). These glomus tumors grow very slowly, and appear late in life. They are thought to arise in early embryology, when our developing brain is twisted in two bends that later disappear, and at a time when 12 cranial nerves start growing out of the brain. The symptoms that appear in cranial nerves 5–12 were mapped. These binary data came from middle-age or older patients who had something wrong with one or more of these nerves. Binary Euclidean distances were calculated from the presence or absence of symptoms. The constructed chart resembles the patterns in early development. Thus it appears that a pattern that was determined during early fetal development can be uncovered by MDS of adult data. Figure 13.18 shows the constructed chart from an MDS analysis of cranial nerve symptoms in adults along with a sketch of the developing nervous system, both showing two 'bends' in similar locations. Constructed cartography can excel at finding patterns in messy medical data, revealing otherwise obscured relationships.

13.3.5 The Future

An exciting extension of the efforts to chart both the spread of diseases and our attempt to fight those diseases is seen in Fig. 13.19. Seen are some 'angry' cancer cells 'looking out upon their opportunities' to find a new home. Their view of the human body is different than ours, thus the anatomical map appears warped to us. Similarly, some attempts to mitigate that disease are illustrated with the falling

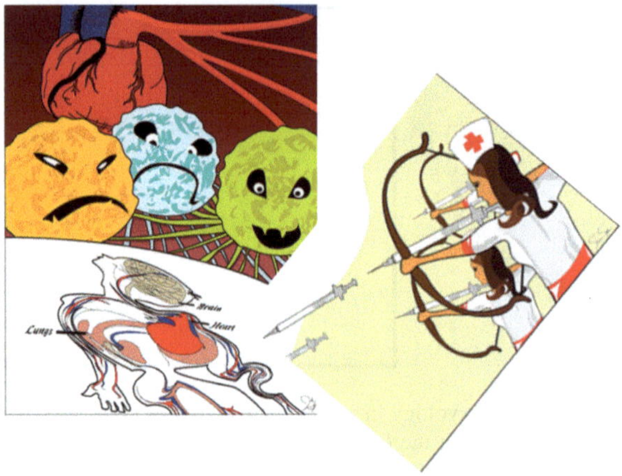

Fig. 13.19 Patterns in how diseases spread and how we attempt to thwart that spread (By JMU Creative Services)

hypodermic needles. This treatment could be chemotherapy agents or surgical excisions. We can only hope that the constructed charts of our efforts to mitigate disease are congruent with the underlying patterns in how the disease spreads. In this way, constructed cartography might help solve on the most 'wicked problems' that we face.

Finally, MDS is increasing being used to map genetic changes. Many people are mapping the genome. Constructed cartography as herein described could be viewed as a way to map the phenotype (or observed behavior) of disease. Perhaps we could call this mapping the 'phenome'. Hopefully some day, charts of both the genome and the phenome will be constructed side-by-side, and any discrepancies therein will be highly instructive. More likely, genetic difference will form meaningful subsets for constructed charts of various diseases. (Any reader with access to data on genetic differences as well as measures of how a disease spread is invited to contact the author and maybe we can collaborate on this important modern wicked problem, predicting successful methods of mitigating the spread of disease from genetic differences.)

13.4 Summary

Constructed cartography can extract underlying patterns in many different forms of 'noisy' data. From an individual's perspective the technique can help identify meaningfully different subsets. That is, it would help an individual know if they were a member of a group (possibly by age, gender, exposure, or genetic background) that might have unique patterns of disease or successful treatment. The technique can help patients understand how their disease spreads in their bodies, in their minds (their knowledge and perceptions of risks), and in their expenses.

For policy makers, the technique can help identify meaningfully different subsets. Constructed charts can identify where one form of effort (such as costs) are not congruent with intended success. The technique can find patterns in how diseases and preventions spread across populations.

From a researcher's standpoint the technique lets underlying patterns in the data emerge without human bias in what the important underlying trends might be. Patterns that emerge can be rigorously evaluated.

13.5 Opportunities to Help Construct Charts of patients' Perceptions

There are several (IRB approved) opportunities to be a voluntary participant in current attempts to construct charts of perceptions. These are Web sites that will record your directly measured proximities.

A. If your child (ages 5–13) is on the autism spectrum disorder or if your child wishes to be a control participant (no ASD), please visit http://csdsquared.cs.jmu.edu/ASDWeb1/Invite.htm, where you will find consent and assent forms, instructions, and simple questions about what small toys and treats are most desired. The test will only take a few minutes and has no more than minimal risk.
B. If you are an adult lay person in VA or MN interested in critical infrastructure or if you are a security expert, please visit http://csdsquared.cs.jmu.edu/InfraStruct/consentform.htm to provide your subjective judgments of infrastructure.
C. If you are an adult lay person, professional artist, physician, medical student, or other health-care professional please visit http://www.uth.tmc.edu/oto/ArtInstruction.htm to provide your subjective judgments of portrait art , and to read more about our studies into the 'art of observation'.

Acknowledgements The following institutions provided important support: Rockingham Memorial Hospital (RMH), Karolinska Hospital, M.D. Anderson Cancer Center (MDA), Cancer Research UK (CRUK), University College London (UCL), University of Liverpool (UK), Johns Hopkins (JHU), Tata Memorial Cancer Center (Mumbai), Aga Khan Med School (Karachi), Fulbright Commission, NASA, Cray Research Inc. The following helped in making maps of costs: Jennifer McCabe & Drs. David Bernstein & Bill Grant, JMU; Dr. Kathleen Philbin, UTx; Dr. Dave Grembi, Dr. Steward Pollock, Kay Harrison, Mike King, Steve Nelson, Deb Thomson, Carrie Willets & Theresa Yoder (RMH); in making maps of head and neck cancers: Drs. Tom Robbins and Bob Byers, MDA; Dr. David Sidransky, JHU; Dr. Iftikar Sallahudin, Aga Khan; Drs. Ashok Mehta & D.N. Rao, Tata; Drs. James Brown and Julia Woolgar, UK; Drs. Randall Morton & Dev Tandon, NZ; in making maps of breast cancers: Drs. Michael Baum & Jayant Vaidya, CRUK; Dr. Thomas Hatschek, Karolinska; Dr. Ragendra Badwe & I. Mittra, Tata; Dr. Tariq Siddique (Aga Khan); in the visualizations: Dr. David Bernstein and Kurt See, JMU; Scott Klasky, Richard Lutowski, Yunsong Zhu, Rongrong Wu, Drs. Dorothea Wiarda, Tarryn Witten, UTx; in making maps of patients' and professionals' perceptions: Drs. Keith Herrin & Charles Stiernberg (UTx), Garland Novosad & Ruth Tornwall (dental hygienists); in maps of migrations: Dr. John Noftsinger,

Ken Newbold & Ben Delp, JMU; in making maps of vaccines: Jennifer McCabe, Drs. David Bernstein & Jerry Benson (JMU). Vincent Vander Poorten MD PhD, made valuable comments on an early draft.

Literature Cited

Breier JI, Fletcher JM, Denton C, Gray LC (2004) Categorical perception of speech stimuli in children at risk for reading difficulty. J Exp Child Psychol 88(2):152–170

Cousan R (2004) Actus Cancer du Sein (article in a Paris magazine about breast-cancer research, including a description of our Web site). La Femme

Gray L (1995) Method of predicting carcinomic metastases, ed. US Patent Office

Gray L (1996) Method of predicting carcinomic metastases, Coninuation in Part, ed. US Patent Office

Gray L (1997) Functional mapping of glomus tumors reveals underlying embryology. Int J Med Inform 44(3):163–168

Gray L (2001) Use of psychophysical scaling for peer-review in problem-based learning. Med Educ 35(11):1089–1090

Gray L, Robbins KT, Byers R (1992) Multidimensional scaling of head and neck metastases. Int J Biomed Comput 31(3–4):177–187

Gray L, Klasky S, Byers R (1993) Visualizing complex patterns in the spread of head and neck cancers. Int J Supercomput Appl 7:167–178

Gray L, Woolgar J, Brown J (2000) A functional map of cervical metastases from oral squamous cell carcinoma. Acta Otolaryngol 120(7):885–890

Gray L, Herrin HK, Stiernberg C, Novosad G, Tornwall R (2002) Perceptions of tongue lesions by dental hygiene students and otolaryngologists. J Cancer Educ 17(4):191–195

Gray LC, Vaidya JS, Baum M, Badwe RA, Mittra I, Siddiqui T, Wiarda D (2004) Functional maps of metastases from breast cancers: proof of the principle that multidimensional scaling can summarize disease progression. World J Surg 28(7):646–651

Gray L, McCabe JA, Bernstein D (2010) Constructed charts of vaccination strategies. Health Informatics J 16(1):25–34

Havey D (2011) Archive for the sphere tag; Tutorial #5: The IP-mapping globe. Accessed 2011

Herrin HK, Gray L (1997) Perceptions of oral cancer risks in smokers and non-smokers. Oral Dis Update 3:6

Kruskal JB, Wish MW (1978) Multidimensional scaling. Sage, London

Noftsinger J,Newbold K, Gray L, Delp B (2008) An innovative approach to studying migration: applying functional mapping to examine global migration trends. Forum on Public Policy. Accessed

Pruzansky S (1983) How to use SINDCAL, a computer program for individual differences in multidimensional scaling. MDS-2 Distribution Tape. Bell Laboratories, Murray Hill

Schiffman SS, Reynolds ML, Young FW (1981) Introduction to multidimensional scaling: theory, methods, and applications. Academic, New York

Chapter 14
"Effectively Living Near the Edge": An Extraordinary Management Tool for Mastering Wicked Problems

Richard N. Knowles

Abstract We live in a world of increasing levels of uncertainty and ambiguity with fewer and fewer tools from ordinary management that are useful in addressing the challenges we all face. Most people live in a world of complexity where the way in which we choose to engage each other and to work is critical. This is the world of wicked problems. The diagram below illustrates this world—showing that the ways to live with the highest possibilities of success are at "The Edge of Chaos"—tche world where Authentic Leadership is the key to a more sustainable future.

14.1 Summary

We live in a world of increasing levels of uncertainty and ambiguity with fewer and fewer tools from ordinary management that are useful in addressing the challenges we all face. Most people live in a world of complexity where the way in which we choose to engage each other and to work is critical. This is the world of wicked problems. The diagram below illustrates this world—showing that the ways to live with the highest possibilities of success are at "The Edge of Chaos"—the world where Authentic Leadership is the key to a more sustainable future.

R.N. Knowles, Ph.D. (✉)
Richard N Knowles and Associates, Inc, St. Petersburg, FL, USA
e-mail: RNKnowles@aol.com

S.F. Kovacic and A. Sousa-Poza (eds.), *Managing and Engineering in Complex Situations*, Topics in Safety, Risk, Reliability and Quality 21, DOI 10.1007/978-94-007-5515-4_14,

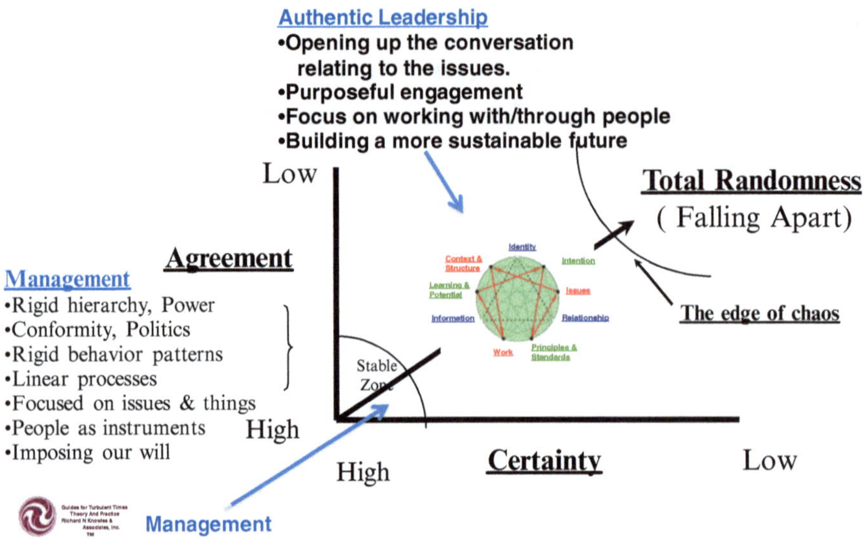

Fig. 14.1 Authentic leadership and management

A powerful tool that enables people work and to live at "The Edge of Chaos" is the Process Enneagram© (the diagram in the center of this figure). The Process Enneagram is the best known tool to simultaneously help people reach rational solutions to complex problems, develop the social connections for the work and build the necessary emotional energy and commitment to solve critical, wicked problems.

The use of this Process Enneagram tool in a highly focused, disciplined dialogue with the people involved with the situation enabling them to develop a map and path forward where they can see both the whole, the parts and the interconnections. What emerges is their living strategic plan and their path-forward to organizational and team coherence, clarity of purpose, authenticity, leadership, purposeful behaviors and (most importantly) *extraordinary results* for the people and the organization.

In using this approach, the author, led the DuPont plant in Belle, WV in reducing injury rates by 98% and emissions to the environment by 88% while increasing productivity by 45% and earnings by 300%. This is described in his book, *The Leadership Dance, Pathways to Extraordinary Organizational Effectiveness ©2002.*

This powerful tool is introduced and illustrated. It has been successfully used in many countries, organizations and situations to address tough issues—the wicked problems that keep resurfacing, recycling and reoccurring. The Process Enneagram is a tool for authentic leaders who want to make a purposeful, positive difference in their workplace and in their lives. This is a leadership tool for building a more sustainable organization.

Through the use of this tool leaders and managers can purposefully engage the natural tendency of self-organization, a powerful force pervading all organizations Knowles (2006).

14.2 Introduction

When a typical manager walks around the facility, he or she often spots people talking together in small, informal groups. They may be talking about a work issue, but sometimes the conversation is about something that may seem totally unrelated to work. What is important is that they are talking about something that interests them. Perhaps it is a work problem, a sports event, a political situation or family problem; maybe they are just getting caught up with each other. Managers often interpret this as a waste of time, so they push the people to get back to work. Pushing people like this wears everyone out. Push, push, push takes a lot of energy, creates friction, and demoralizes everyone. It is a huge waste of time and energy. And it is not a sustainable way to lead.

What is happening as people gather to talk? What is going on here? Why? What can we learn from this? Is it just a waste, or is there something deeper here that could be a key to opening up the energy and creativity of the organization? What would an organization be like if everyone were working to his or her best, applying themselves to doing what it takes for the business to succeed? What would it be like if each person were working on an opportunity like improving workplace safety while lowering costs or simultaneously improving productivity, quality, and customer service? What would it be like if they were mastering wicked problems?

There is something in this way people come together that is extremely important and, if a leader chooses to engage with it, can open up the effectiveness, efficiency, and productivity of the organization by 30–40%.

This phenomenon relates to the concept of self-organization, which can sound like chaos, anarchy, and potential failure. When the author first heard of this, about 15 years ago, he almost fell out of his chair. After all, he was the plant manager and was supposed to keep things organized and tight. But over the years, in learning how to engage with this natural tendency of self-organization, he found that in persistently talking with and listening to all the people about important issues like safety, quality, costs, and sustainability, the enterprise's impact on the community and the quality of work life significantly improved the organization's performance. Injuries dropped by 98%, productivity rose 45%, emissions dropped 88%, and earnings rose 300%. Together, all of us confronted and struggled with the issues and developed coherence and clarity about what we were trying to do and why this was important to all of us. This was done with openness, honesty, authenticity and hard work.

14.3 Engaging the Natural Tendency of Self-Organization

There is a natural, pervasive tendency for all living systems to "self-organize." This tendency can be seen throughout nature at all levels, from tiny bacteria to large ecosystems. In this essay, a system is loosely defined here as a collection of similar things, a group or organization. Since the parts of the system are similar in some way, there is a shared identity that defines a sort of boundary around this collection of things.

People experience this self-organizing tendency when they gather together to talk, work, and play. This phenomenon is so pervasive and subtle that it's usually not even noticed. Yet it is happening all the time, as people self-organize around something that is important to them.

This natural tendency is powerful, yet subtle; it is like the current in a flowing river. Often people join the flow and engage purposefully with this tendency in their conversations, in informal gatherings like family reunions, or in high-performance work teams. However many managers have worked counter to this by trying to impose their wills on people, using a command-and-control approach, to achieve a specific task or to reach a specific goal. This is non-purposeful engagement with the natural tendency of self-organization. As leaders and managers strive to get the job done, to complete the task, or to reach the goal, they bump up against this tendency to self-organize. Using the command-and-control approach is like trying to take the twists and turns out of a river and make it flow the way we want. But by purposefully engaging this natural tendency, leaders and managers can join the river and draw great energy and focus from it. Self-organization exists in organizations of all sorts all the time.

Much of the vast literature on management and leadership is directed at ways that one's will can be imposed on the people in the organization to accomplish the tasks at hand. Most managers crave stability, reliability, predictability, and control in their organizations. While imposing conditions like these is necessary for a machine like an airplane, this approach suppresses the purposeful vitality, energy, and creativity of people. Imposing one's will becomes command-and-control management when it is pushed too far. When this imposition mode of managing and leading is used in an enterprise, people self-organize in ways that are seen by the organization as non-purposeful, becoming lethargic, unresponsive, and resistant to change. Such organizations behave as if they must be pushed and shoved if they are to do what management wants them to do. These organizations behave as if they are unhealthy living systems: torpid and passive. There is a growing frustration with this way of leading because of the less-than-hoped-for results, the effort required to keep things moving, the lack of sustainability, and the negative, self-organizing behavior that it generates in people.

Since the early 1960s, much has been written about teams and how some of them can accomplish extraordinary things. Reading between the lines, we see that many of the teams that do best are the ones that have learned to purposefully engage the natural tendency for people to self-organize. They are called "high-performing work teams" because they accomplish so much.

14.4 Our Leadership Choice

So, leaders and managers always have a *choice* to make about the way they engage this natural tendency for people to self-organize. They can find ways to engage this tendency purposefully or they can try to non-purposefully impose their wills. There are times when the situation is such that one of these choices may be more appropriate than the other. However, if leaders and managers can purposefully engage this natural tendency most of the time, they will be in the most sustainable position. This is not about "good" or "bad," but rather, it's about choosing the most effective way to lead in a particular situation, at a particular time. Leadership is very much a temporal process in which the leader must be conscious of what is happening and must choose the most appropriate leadership process for the situation; this is the Leadership Dance.

Most managers have learned how to use command-and-control management and leadership processes, but only a few have learned how to use management and leadership processes that purposefully engage the natural tendency of people to self-organize. Often, this is based on the intuitive sense of these leaders who know that the command-and-control processes aren't very effective over the long term.

Increasingly, a language and models that are useful in working purposefully with this natural tendency of self-organization are surfacing (Knowles 2002). Combining powerful models and explicit terminology with intuitive insights provides an effective way to purposefully engage the natural tendency of self-organization. I call management and leadership processes that purposefully engage the natural tendency to self-organize "Self-Organizing Leadership ™"; it is authentic leadership. When the leader purposefully engages the tendency of self-organization, vitality, energy, and creativity increase, and the organizations behave as if they are healthy living systems. The fundamental idea speaks to the nature of relationships as they are developed and expressed in conversations. Stacey (2001) is leading explorations into the importance of conversations in organizations in his work on complex responsive processes (CRP).

Guastello et al. (2009), *Chaos and Complexity in Psychology, The Theory of Nonlinear Dynamical Systems*, have offered specific propositions regarding self-organization theory which fully support the approach of Knowles (2002) in this chapter.

Much good work is being done to help understand the deeper principles behind the way self-organization unfolds. These theoretical foundations are critical to building a solid groundwork for this important work for leaders. Complex responsive processes look at the conversations among the people in the organization as temporal events. With the CRP approach, leaders have direct engagement with people and are not separated from what is currently taking place in the organization. On the other hand, the theory of complex adaptive systems (CAS) looks at systems and organizations as things to be acted upon. With the CAS approach, the engagement is with the people in the organization as if they are an object that is different from the leader. Both CRP and CAS approaches are useful in helping to develop deeper insights into what is happening in organizations, providing that the distinction between the two methods is understood and made explicit.

Interestingly, to purposefully engage the natural tendency of self-organization, all we need to do is start having the important, often intense, sometimes difficult conversations about the critical issues, the wicked problems, facing us and invite others to join in the exploration. Three areas provide important conversational pathways that allow us as leaders to purposefully engage the process of self-organization. These are the fundamental pathways for Self-Organizing Leadership:

1. Abundantly sharing important, relevant information, such as various aspects of how the organization is performing—the competitive situation, the costs, environmental and societal impacts of what we are doing, safety issues, quality and

productivity concerns, earnings, and the potential impact of all this on our organization's future;
2. Building interdependent relationships and trust by spending time with people on their turf, listening and sharing ideas, keeping our word, taking public responsibility for our mistakes, and talking together about how we'll correct the situation; and
3. Helping people find how they and their work fit into the whole picture, helping them to see the positive impact of their work—discovering meaning in the work.

Authentic conversations, one person at a time, begin to open up the connections that are the medium of successful self-organization. Leaders and managers play the key role in creating the conditions for this to happen.

These authentic conversations must be about the questions and issues that are truly important; they must be critical to the success of the work and the goals of the organization. These conversations may be difficult, so it requires courage, concern, commitment, and care to stay in the "heat" and find new ground on which we can build. We have to be open, honest, and transparent as we do this together. *If transformation is to occur, we need to be engaged within the processes of the organization* and *not acting on the organization as if it were an external thing.*

There are a number of ways to open up these conversations. We can begin to share important, relevant information, building trust and meaning as we talk and work together. We can engage in dialogue processes. We can ask questions about what we see or sense and ask why. We can use story-telling as a way for people to find meaning in what is happening. Margaret J. Wheatley (1992) was one of the early thinkers to open up this way of leading. We can use the Open Space Technology of Harrison Owen (1991) to explore people's interests in a particular subject. We can use the Future Search approach of Sandra Janoff and Marvin R. Weisbord (1995) to find out what is important to people and who in the organization cares enough to carry it forward. David Cooperrider's Appreciative Inquiry (2005) approach is also a great way to open up the conversation in a positive way. Sometimes it is about having the hard conversations that Susan Scott talks about in *Fierce Conversations* (2004). Sometimes it is using Glenda Eoyang's (Eoyang et al. 2001) approach to explore the difference that makes the difference. The challenge is to keep the conversations open, flowing, and authentic over time.

Most people find that developing relationships by listening and talking together is stimulating and exciting. It is the first time many have been heard or taken seriously in their entire career. New ideas are shared, exciting possibilities are discovered, and opportunities may open up for significant improvement. It's important to document the conversation, to keep the conversational space open, to keep the conversation alive carrying it forward to others who need to be engaged. It is important to simply and effectively document the critical questions and issues raised in the conversation so that the "space" can be held open to carry these conversations forward with those who may not have been initially involved.

The approach presented in this chapter is the applications side of the theories of Jeffrey Goldstein, James K. Hazy, and Benyamin B. Lichtenstein as expressed in their books *Complex Systems Leadership Theory* (2007) and *Complexity and the Nexus of Leadership* (2010).

It is also consistent with the Joseph C. Rost (1991) who defined leadership as an influence relationship among leaders and followers. This approach to leadership is key to building a more sustainable future. Jeana Wirtengerg (2009) describes this more fully in Knowles, Chapter 1, *Leadership for Sustainable Enterprise, The Sustainable Enterprise Fieldbook.*

14.5 The Process Enneagram©

One way to effectively address and document the critical questions and issues is through a cyclical, disciplined, focused progression of discussions that develop successively deeper and more coherent insights. In his experience in working in organizations, the author has found that most of the information an organization needs to accomplish its work is already scattered among the various individuals within the organization. This open, honest progression of conversations provides a way to develop a shared understanding of all they know. A path for transforming the organization and ourselves opens up as they move forward in their journey toward completeness.

As mentioned earlier, the Process Enneagram is the best known tool to simultaneously help people reach rational solutions to complex problems, develop the social connections for the work and build the necessary emotional energy and commitment to solve critical problems.

This cyclical progression of conversations can be easily mapped onto a Process Enneagram map (Knowles 2002) to capture the ideas, to keep open the space for future conversations and to develop a living strategic plan. The Process Enneagram is a fundamental pattern for the deep processes of Self-Organizing Leadership. It can be used in any organization, for any problem at any level of scale; it is fractal in nature. A. G. E. Blake (1996) has written deeply and extensively about the theory of the Enneagram of Process; this work is entirely consistent with the practical applications of the work presented here.

The progression begins with a conversation intended to collectively define a clear, compelling question, a wicked problem, or challenge that a group or organization is facing. Then it moves on to having a focused conversation about who they are and their history, both as individuals and as an organization; it is about their Identity. The process then shifts to defining the organization's Intention so that they can develop a shared, co-created picture of just what they want to have happen in solving the wicked problem before them. They then co-create shared Principles and Standards of behavior and explore the dynamics of how these co-created Principles and Standards of behavior will enable them to work together more effectively as they work towards fulfilling their Intention. Co-created Principles and Standards profoundly affect their relationships as well as many of the issues they have already identified. The progression then moves on to conversations about the Issues and

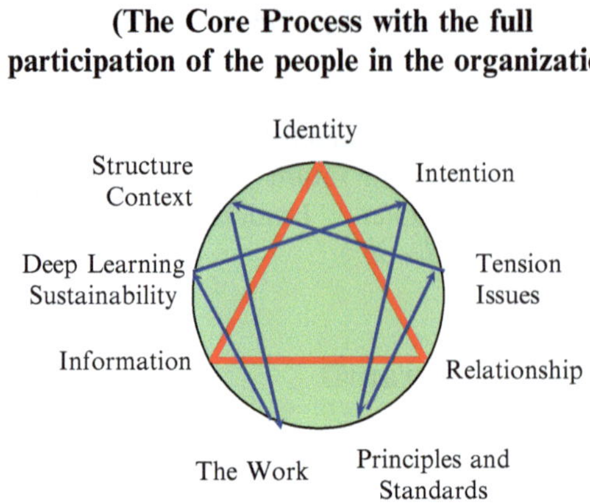

Fig. 14.2 The process enneagram

Tensions facing them; the barriers they must overcome or learn to live with like the either/or ambiguity of production and safety. As they work together on the Issues, applying their Principles and Standards, their Relationship develops trust and interdependence. They next look at how to best Structure and organize themselves to accomplish the Work that needs to be accomplished. At this point in the process they also look at the external Context, the environment in which they are working; this provides them with an external coupling to the outside world. The specific tasks and Work they'll do are the next focus. All the Work must advance them towards fulfilling the Intention they have identified. Then they look at how they will continue to share meaningful Information, Learn and Grow and discover new potential for their future together. As they carry forward together into more and more cycles and widen the conversation, other insights will emerge that can be added to the map they are creating. In this cyclical process, they move up a spiral of learning and growth. This cyclical progression of focused, disciplined conversations enables the development of a very high level of coherence, purposefulness, sustainability, and will for action. These are the critical elements for Mastering Wicked Problems.

A second process is running beneath this one just described. It relates to freely sharing information, building trust and interdependent relationships and creating their identity. This is the process of Self-Organizing Leadership©. These emerge during the open, honest conversations described in the preceding paragraph.

The Process Enneagram diagram is illustrated below. It clearly shows the nine interacting points, their relationships and the progression through them. As the conversations proceed, key ideas are gathered at each point. A map of their strategic plan emerges in these conversations. These ideas are gathered as a collage at each point. The map is kept constantly before them as a guide and is changed as things progress; it is a living strategic plan to address the wicked problem.

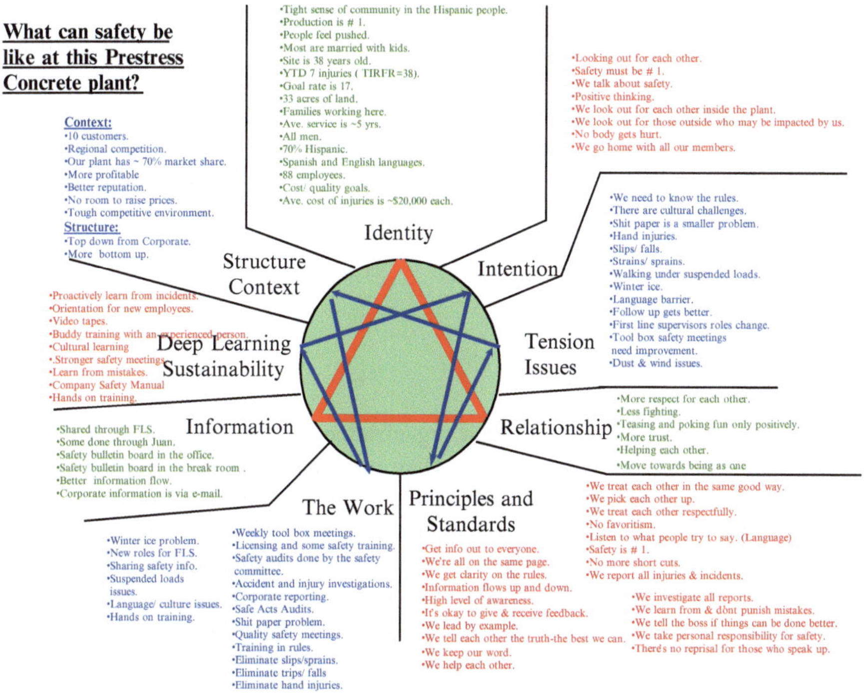

Fig. 14.3 Sample process enneagram application

A real-life example of a Process Enneagram map developed at a prestressed concrete beam plant is show below. The workers at this plant had been suffering very poor safety performance with a marginally successful top-down management approach to improving safety. As the map was developed as part of a 2-day workshop, the people transformed the way in which they saw safety and their role in making significant improvements. As a result of their new understanding and commitment, their injury frequency dropped from 1 serious injury a month among the 80 men to having none for 10 months. They were still using this map as their guide 4 years later and have sustained the improved levels of injury frequency rates as well as experiencing less severity when they do have an injury. Because of the way they have decided to work together, this approach has broadened into their other work areas and they are seeing improved costs and quality.

In creating their Process Enneagram map together, control shifts from management edicts and pronouncements to the co-creation of "the Bowl" (Knowles 2002, p. 99). The Bowl consists of the mission, vision, expectations, principles, and standards of performance within the organization or unit, all collected on the map. The map is posted and discussed in all meetings. The Bowl is the container for the organization's work and provides both order and focus for the organization. Within the Bowl, people work in a self-organizing way with a high level of freedom to accomplish the tasks before them. In this open, energetic, creative space wicked problems are mastered. The Bowl provides the container, the order and focus, so that they don't go all over the place which can result in the organization losing focus.

As leaders purposefully engage the natural tendency of self-organization in this way, the energy and creativity of people flows and the effectiveness of the organization improves significantly—often more than 30–40 %. Resistance to change almost disappears.

14.6 Emergence

As leaders engage the people in their organizations and the mode of engagement shifts, new types of behaviors emerge. Here we examine some of these. As mentioned earlier, leadership is very much a temporal process, and each moment is new.

The author describes behaviors emerging from three different leading processes in *The Leadership Dance, Pathways to Extraordinary Organizational Effectiveness* (2002, pp. 169–176). These leading processes are embedded in the Process Enneagram and each consists of three interdependent ideas (Knowles 2002, p. 30). In actual practice, all of these are running all the time, but it useful for this analysis to look at them as if they were separated.

The most basic leadership process is the Self-Organizing Leadership process, which consists of the interaction of identity, relationship, and information.

Identity, relationship, and information emerge as we, all of us in the organization, engage in dialogue about questions and issues that are very important to us, that is, when we engage with the natural tendency of self-organization in purposeful ways. Reflecting on the importance of these conditions for self-organization, we can look at them from the perspective of their threefold relationship. They can be seen and experienced as forces that are interacting all the time. In their interaction new behaviors emerge:

- When we, all of us together, have a clear sense of our identity and an interdependent relationship, as new information becomes available we can move into *action*;
- When we have an interdependent relationship and an abundance of information, as we become more clear on our identity *meaning* emerges;
- When we have a clear sense of our identity and an abundance of information, as our relationships become more interdependent *trust* emerges.

Action, meaning, and trust are critical to releasing energy and creativity and making extraordinary things happen. Emergence in the Self-organizing Leadership Process. (Knowles 2002, p. 130)

A second leadership process embedded in the Process Enneagram is Operational Leadership. This process relates to the issues all of us are facing, the organizational structures we create to try to address these issues, and the work that needs to be done to solve the issues. From time to time, every manager and leader needs to use this process. But, if this leadership process is overused, it becomes the command-and-control

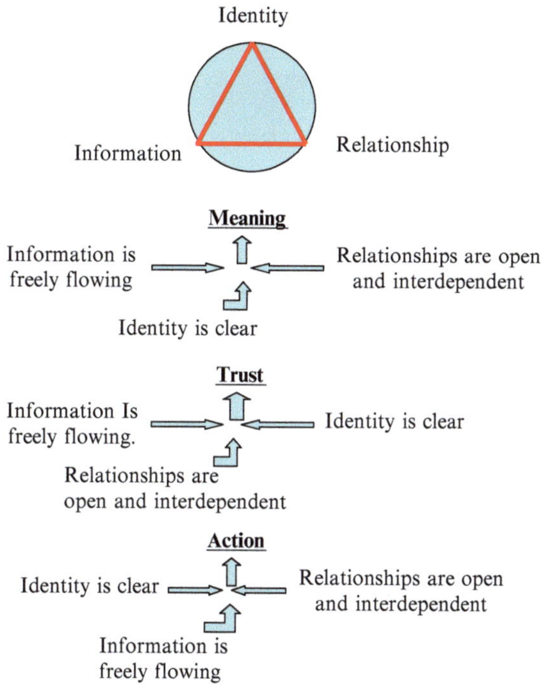

Fig. 14.4 Emergence in the self-organizing leadership process

process as things are imposed from the top and driven down into the organization; this is the non-purposeful engagement with the natural tendency of self-organization.

When the leaders of the organization decide to move into the Operational Leadership mode and there has been clarity and coherence developed in the dialogue around the Process Enneagram, people are involved and know what is going on, and the following may occur:

- When the work is focused on fulfilling the intention and the issues have been examined, as the structure becomes more self-organizing a *sense of urgency* emerges;
- When the structure is self-organizing and the work is focused on fulfilling the intention, as the issues are examined a *clarity of purpose* emerges;
- When the structure is self-organizing and the issues have been examined, as the work is focused on fulfilling the intention, a *sense of purposefulness* emerges.

When the leaders of the organization move into the Operational Leadership mode and engage with the natural tendency of self-organization in non-purposeful ways and where the level of incoherence is high because people aren't involved and don't know what is going on, the following may occur:

- When the work is unfocused and the issues are unexamined, as structure is imposed *fear* and *anxiety* emerge in questions like, Why are we reorganizing again? and, What's going to happen to me?
- When the structure is imposed and the work is unfocused, as they address the unexamined issues *confusion* emerges in questions like, Why are we doing this

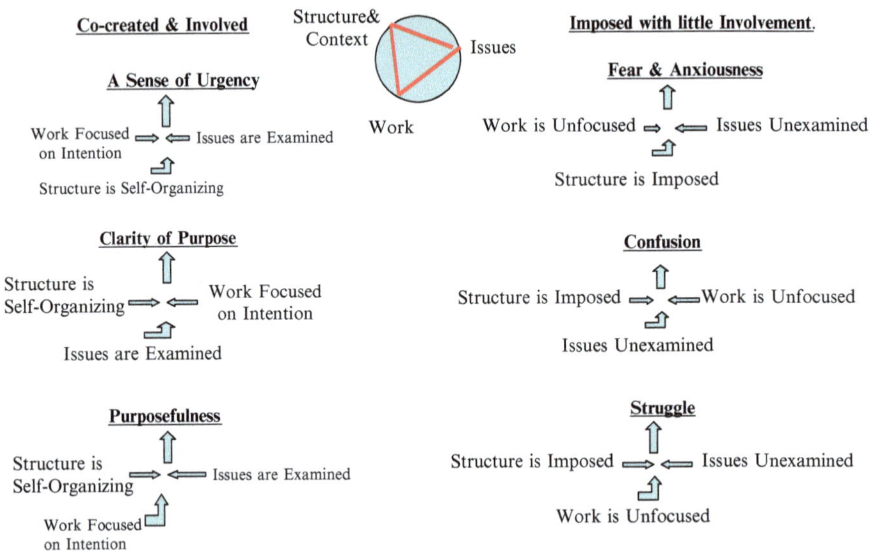

Fig. 14.5 Emergence in the operational leadership process

when we have this other, more important problem? and, Why doesn't management listen to us?

- When the structure is imposed and the issues are unexamined, as they engage in imposed and unfocused work *struggle* emerges in questions like, Why do I have to do this now? and I've learned to do the other job and now you're making me do this new one. Why?

Depending on the leader's choice in engaging the natural tendency of self-organization, vastly different outcomes are experienced.

A third leadership process that is embedded within the Process Enneagram is the Strategic Leadership Process. This process relates to the organization's intention (what it wants to do), its principles and standards of behavior that are needed to support the intention, and the learning and potential that can come out of the new things the people in the organization need to do.

Depending on the choice of the leader's mode of engagement with the natural tendency of self-organization, very different outcomes emerge.

When the organization transitions into the Strategic Leadership Process and the leaders have purposefully engaged the people in the organization so the level of clarity and coherence are high, people are involved and know what is happening within the organization, or their unit, so the following may occur:

- When the people are open to learning and the intentions are clear, as they engage in co-creating the new principles and standards, *hope* emerges;

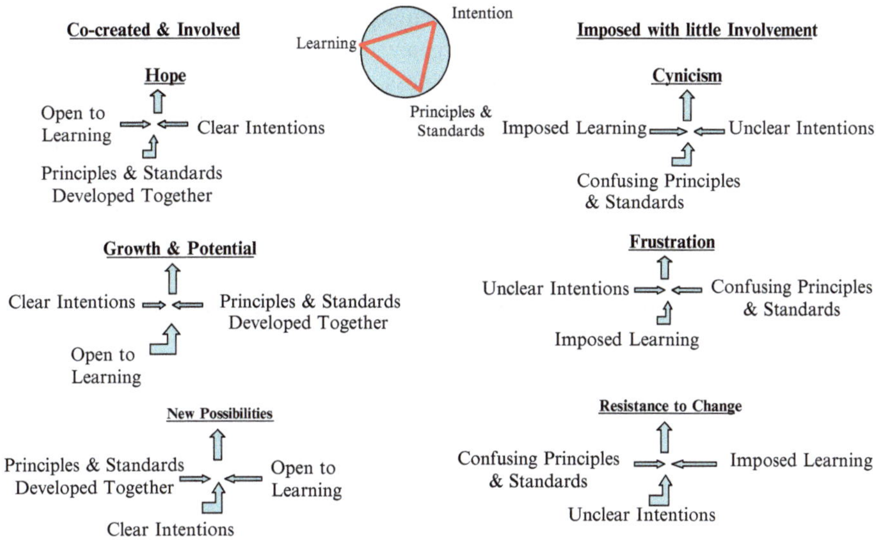

Fig. 14.6 Emergence in the strategic leadership process

- When the people have clear intentions and their principles and standards have been co-created, as they are open to new learning, *growth and potential* emerge;
- When the people have co-created principles and standards and are open to learning, as their intentions become clearer, *new possibilities* emerge.

When the leaders of the organization decide to move into the Strategic Leadership Process and have engaged the natural tendency of self-organization in non-purposeful ways in which the levels of incoherence are high, people aren't involved and don't know what is happening within the organization, the following may occur:

- When the learning is imposed and the intentions are unclear, as confusing principles and standards are imposed, *cynicism* emerges in questions like, Why are you doing this to me when our corporate values say we value people? and, Why should we believe management now when they have not kept their word in the past?
- When the intentions are unclear and they have confusing principles and standards, as the learning is imposed, *frustration* emerges in questions like, Why do we have to learn this new stuff when what we've done in the past was good enough? or You mean that I'm going to have to come in on overtime and find child-care support to learn this stuff?
- When they have confusing principles and standards and the learning is imposed, as they address unclear intentions *resistance to change* emerges in questions like, This is just the flavor of the month; why bother? and, Is this just another change to get rid of my job?

Therefore, if leaders purposefully engage with the natural tendency of self-organization, the people in their organization will tend to exhibit a sense of urgency, clarity of purpose, purposefulness, hope, growth and potential, and openness to new possibilities. This is the most likely environment where the wicked problems can be successfully addressed. In this environment, resistance to change almost disappears.

On the other hand, if leaders choose to engage with the natural tendency of self-organization in non-purposeful ways, the people in their organization will tend to exhibit fear and anxiety, confusion, struggle, cynicism, frustration, and resistance to change. This is an environment where the challenge of wicked problems will be too much for the people.

The choice is simple, but the execution can be difficult.

14.6.1 Some Examples

The first example relates to a crisis like a fire, power outage, or major storm. When there was a fire in the DuPont chemical plant in Belle, W.V., where the author was the plant manager, it was interesting to see how people behaved. The highly trained fire brigades, operators, mechanics and supervisors self-organized around getting the fire out, cleaning up, and getting back into production. People did extraordinary work together in teams that spontaneously formed and re-formed as the tasks changed. There was a lot of energy and creativity in all they were doing. The managers were so busy with all of their tasks that they were not able to micro-manage everything, so the people on the floor quite successfully stepped forward to do what was needed. However, once the crisis was over, things reverted to the normal command-and-control management approach. Energy dropped, and creativity moved back underground.

You may have had this experience yourselves in those times when you were living through a crisis. There probably was a lot of self-organization going on during the big hurricanes in 2005, but what was reported related to the problems and frustrations imposed in the command-and-control efforts used by the federal, state, and local governments.

Another example from the time when the author was plant manager at the Belle plant will help illustrate these ideas further. From the very beginning of the construction project to convert from pneumatic to electronic process control systems at Belle, the engineers, operators, mechanics, and all supervision were deeply involved in the communications and planning processes. The goal was to convert to the new chemical process control systems without maintaining the old systems for transition and backup. This was a high-risk approach for if we failed, we'd go out of the business. Everyone needed to be involved in the weekly project status reviews, planning sessions, design meetings, and the like; many of the operators, mechanics, and engineers were sent to the Honeywell School for computer training. All the information was shared on a continuous basis, interdependent relationships were developed and everyone knew how they fit into the larger picture and the critical need for their contributions for a successful project. There was a lot of give-and-take in these meetings as everyone tried his or her best to make the project a success. At the end of the project, the unit was started up without incident making high quality product in record time. This approach cut the costs and time in half, from the original estimate of $6 million in investment and 2 years to implement. Then 15 more projects were successfully put into place in other operations in record time and at lower-than-forecast investment without running any parallel processes, clearly showing the success of self-organizing leadership processes.

In a third example the author used engagement processes like these, with the City of Niagara Falls, N.Y., Leadership Team who worked together with the mayor in a way that resulted in cutting out $15 million from a $62 million budget over a 4-year period. This was the first time in the city's history that the Leadership Team worked together this way, and saved so much money. Sharing information, building interdependent relationships, and getting very clear on the mission to make the city as strong as possible were keys in this success. A lot of the savings came about as people talked about what was going on, so, for example, we knew to put the new sewers into place before paving the streets.

Surely, most of you reading this paper can think of examples in which well-intended projects with high expectations were started from the top of the organization with little employee involvement. The people resisted the changes, slowing things down to the point that the organization lost energy and interest, finally giving up on the project. Many of the quality improvement efforts over the last 20 or more years have ended like this; however, it's not the poor quality of the technology, but rather the lack of deep involvement of all the people. The poor performance of so many Organizational Re-Engineering projects also relates to this lack of purposeful engagement with the natural tendency of self-organization.

14.7 Application Across Cultures

The Self-Organizing Leadership processes described in this paper have been used across many cultures for more than 14 years. For example, Tim Dalmau (www.dalmau.com) has used this in companies and communities in Australia, New Zealand, South Africa, Namibia, Thailand, United States, Mexico, Malaysia, Germany, Indonesia, China, and Singapore. Steve Zuieback (www.stevezuieback.com) has used this extensively in the state of California school system. The author has used it extensively in Australia, New Zealand, United States, Canada, and the United Kingdom in organizations ranging from heavy industry—like steel, coal, and chemicals—to school districts, accounting firms, the United Way of Niagara Falls, city government, and various community projects like the Niagara County Study on Services for the Aging. Gwen Andrews recently used these processes in her work as assistant principal of the EASY English language School in Weihei, China.

Claire Knowles (www.lightsonworkshop.com) uses this approach very effectively in her work coaching women in transition.

This approach deals with the deeper issues of how people choose to think, to come together, to work, and to be. It is not limited to any particular sort of work, culture, or organization. It applies to situations ranging from individuals to very large groups.

There seems to be an archetypal nature to this work that makes it useful and transferable. While the specific situations differ in each instance, the deeper patterns and processes are highly consistent, can be "seen," understood, and applied.

14.8 Conclusion

As leaders, we have a choice to make about how we encourage and engage the natural tendency to self-organize. While historically leaders and managers have tried to impose their wills and have resisted this tendency—there will still be a few occasions when leaders still need to do this—we are finding that purposefully engaging the natural tendency to self-organize produces vital, coherent, energetic, creative, highly effective, and more-sustainable organizations. Self-Organizing Leadership provides pathways for leaders to effectively and purposefully engage the natural tendency of self-organization.

This authentic leadership work requires a high level of openness, integrity, courage, and commitment. Listening deeply to learn from each other, asking the tough, deeper questions about what is happening, and respecting and truly valuing each other while keeping our eyes on fulfilling the organization's vision, mission, and intentions requires the leader to be working from a deep sense of self, purpose, and integrity. This sustainable way of leading is more about our State of Being than about having a set of skills, as important as they are.

14.9 Making It Personal

The journey into leading this way for the author was long and difficult. In *The Leadership Dance* (Knowles 2002), this journey is described in the Foreword. It is a story to help show that we can all make this journey if we so chose.

The satisfaction in having made this journey is immense. The author moved from being a lonely, tough, top-down manager to becoming a cheerleader for all the good work that people were doing. There were plenty of tough decisions to make, and standards that had to be met, yet in learning together, the transformation was wonderful.

The author learned that highly productive, effective, meaningful organizations can be created in which people can work together in purposeful ways. We don't have to push and coerce people to do the right things and to change. In co-creating the future, resistance to change almost disappeared and terrific results emerged. As coherence emerged, it all came together.

References

Blake AGE (1996) The intelligent enneagram. Shambhala, Boston/London

Cooperrider DL, Whitney D, Stavros JM (2005) Appreciative inquiry handbook. Berrett-Koehler, San Francisco

Eoyang GH, Olsen EE, Beckhand R, Vail P (2001) Facilitating organizational change: lessons from complexity science. Jossey-Bass/Pfeiffer, New York

Goldstein J, Hazy JK, Lichtenstein BB (2010) Complexity and the nexus of leadership. Palgrave MacMillan, New York

Guastello SJ, Koopmans M, Pincus D (2009) Coherence, complexity and information flow: self-organizing processes in psychotherapy. In: Pincus D (ed) Chaos and complexity in psychology, the theory of nonlinear dynamical systems. Cambridge University Press, Cambridge, p 337

Hazy JK, Goldstein JA, Lichtenstein BB (2007) Complex systems leadership theory. ISCE Publishing, Mansfield

Knowles RN (2002) The leadership dance, pathways to extraordinary organizational effectiveness. Center for Self-organizing Leadership, Niagara Falls

Knowles RN (2006) Engaging the natural tendency of self-organization. Transformation. World Business Academy. 20(15):10 Aug 2006

Owen H (1991) Open space technology, a user's guide. Berrett-Koehler, San Francisco

Rost JC (1991) Leadership for the twenty-first century. Praeger Publishers, Westport, p 102

Scott S (2004) Fierce conversations: achieving success at work and in life, one conversation at a time. The Berkeley Publishing Group, New York

Stacey RD (2001) Complex responsive processes in organizations. Routledge, London/New York

Weisbord MR, Janoff S (1995) Future search. Berrett-Koehler, San Francisco

Wheatley MJ (1992) Leadership and the new science, learning about organization from an orderly universe. Berrett-Koehler, San Francisco

Wirtenberg J (2009) The sustainable enterprise fieldbook. Greenleaf Publishing/AMACOM, London/New York

MIX
Papier aus verantwortungsvollen Quellen
Paper from responsible sources
FSC® C105338

If you have any concerns about our products,
you can contact us on
ProductSafety@springernature.com

In case Publisher is established outside the EU,
the EU authorized representative is:
Springer Nature Customer Service Center GmbH
Europaplatz 3, 69115 Heidelberg, Germany

Printed by Libri Plureos GmbH
in Hamburg, Germany